T0212226

Joint Source
Channel Coding
Using Arithmetic Codes

Synthesis Lectures on Communications

Editor
William Tranter, *Virginia Tech*

Joint Source Channel Coding Using Arithmetic Codes
Dongsheng Bi, Michael W. Hoffman, and Khalid Sayood
2009

Fundamentals of Spread Spectrum Modulation
Rodger E. Ziemer
2007

Code Division Multiple Access(CDMA)
R. Michael Buehrer
2006

Game Theory for Wireless Engineers
Allen B. MacKenzie, Luiz A. DaSilva
2006

Joint Source Channel Coding Using Arithmetic Codes

Dongsheng Bi, Michael W. Hoffman, and Khalid Sayood

ISBN:978-3-031-00547-3   paperback
ISBN:978-3-031-01675-2   ebook

DOI 10.1007/978-3-031-01675-2

A Publication in the Springer series
SYNTHESIS LECTURES ON COMMUNICATIONS

Lecture #4
Series Editor: William Tranter, *Virginia Tech*
Series ISSN
Synthesis Lectures on Communications
Print 1932-1244   Electronic 1932-1708

Joint Source Channel Coding Using Arithmetic Codes

Dongsheng Bi, Michael W. Hoffman, and Khalid Sayood
University of Nebraska

SYNTHESIS LECTURES ON COMMUNICATIONS #4

ABSTRACT

Based on the encoding process, arithmetic codes can be viewed as tree codes and current proposals for decoding arithmetic codes with forbidden symbols belong to sequential decoding algorithms and their variants. In this monograph, we propose a new way of looking at arithmetic codes with forbidden symbols. If a limit is imposed on the maximum value of a key parameter in the encoder, this modified arithmetic encoder can also be modeled as a finite state machine and the code generated can be treated as a variable-length trellis code. The number of states used can be reduced and techniques used for decoding convolutional codes, such as the list Viterbi decoding algorithm, can be applied directly on the trellis.

The finite state machine interpretation can be easily migrated to Markov source case. We can encode Markov sources without considering the conditional probabilities, while using the list Viterbi decoding algorithm which utilizes the conditional probabilities. We can also use context-based arithmetic coding to exploit the conditional probabilities of the Markov source and apply a finite state machine interpretation to this problem.

The finite state machine interpretation also allows us to more systematically understand arithmetic codes with forbidden symbols. It allows us to find the partial distance spectrum of arithmetic codes with forbidden symbols. We also propose arithmetic codes with memories which use high memory but low implementation precision arithmetic codes. The low implementation precision results in a state machine with less complexity. The introduced input memories allow us to switch the probability functions used for arithmetic coding. Combining these two methods give us a huge parameter space of the arithmetic codes with forbidden symbols. Hence we can choose codes with better distance properties while maintaining the encoding efficiency and decoding complexity. A construction and search method is proposed and simulation results show that we can achieve a similar performance as turbo codes when we apply this approach to rate 2/3 arithmetic codes.

KEYWORDS

source coding, joing source channel coding, arithmetic coding, channel coding, digital communications

Contents

1 **Introduction** 1
  1.1 Introduction 1
  1.2 Joint source and channel coding schemes 3
  1.3 Joint source and channel coding with Arithmetic codes 5

2 **Arithmetic Codes** 9
  2.1 Encoding and decoding processes 9
  2.2 Integer implementation of encoding and decoding with renormalization 11
    2.2.1 Encoding with integer arithmetic   12
    2.2.2 Decoding with integer arithmetic   14
    2.2.3 Overflow and underflow problems   15
  2.3 Optimality of arithmetic coding 17
    2.3.1 Arithmetic codes are prefix codes   18
    2.3.2 Efficiency   19
    2.3.3 Efficiency of the integer implementation   19

3 Arithmetic Codes with Forbidden Symbols 21
  3.1 Error detection and correction using arithmetic codes 21
    3.1.1 Reserved probability space and code rate   21
    3.1.2 Error detection capability   24
    3.1.3 Error correction with arithmetic codes   25
  3.2 Viewing arithmetic codes as fixed trellis codes 27
    3.2.1 Encoding   30
    3.2.2 Decoding   30
  3.3 Simulations with an iid source 32
  3.4 Simulations with Markov sources 35

3.4.1 Comparing scenario (*a*) and (*b*)   37

3.4.2 Comparing scenario (*b*) and (*c*)   38

4   Distance Property and Code Construction 41

4.1   Distance Property of Arithmetic codes 41

4.1.1 Bound on error events   41

4.1.2 Using the bound to get estimate of error probability   43

4.1.3 Determining the multiplicity $A_{m,l}$ 43

4.2   Verification 44

4.3   Complexity factors and freedom in the code design 45

4.3.1 Complexity factors   45

4.3.2 Freedom in the code design   49

4.4   Arithmetic codes with input memory 49

4.4.1 Memory one arithmetic codes with forbidden symbols   51

4.4.2 Memory two arithmetic codes with forbidden symbols   52

4.4.3 Memory three arithmetic codes with forbidden symbols   52

5   Conclusion 57

Bibliography 61

CHAPTER 1

Introduction

1.1   INTRODUCTION

As vector quantization is the quintessential lossy compression technique, arithmetic coding is the quintessential lossless compression method. Its centrality to compression can be inferred from the fact that the method makes its first appearance, albeit in a form unsuitable for implementation, as part of the proof of The Fundamental Theorem for a Noiseless Channel in Shannon's landmark *A Mathematical Theory of Communication* [92]. The idea as proposed by Shannon was very simple. Given a set of messages of length N partition with probabilities $p_1, p_2, \ldots, p_n$, the code for the sth message is simply the binary representation of the cumulative probability P_s truncated to m_s bits, where

$$\log_2 \frac{1}{p_s} \leq m_s < 1 + \log_2 \frac{1}{p_s}$$

Essentially, this means that given the set of all messages of length N, each message i can be represented uniquely by a number from an interval of size p_i. Furthermore, Shannon showed that such a code would achieve entropy. Theoretically, this code is fine, but Shannon did not provide an implementation. Peter Elias in several unpublished notes developed the idea. The first practical implementations were due to Jorma Rissanen at IBM [79] and Richard Pasco [73] in his doctoral dissertation. Things took off from there - after a brief wait for the IBM patents to expire - and now some variation of arithmetic coding is part of most compression standards.

The efficiency of arithmetic coding also makes it very vulnerable to errors. An arithmetic code is basically a very high resolution binary representation of a very small interval on the real number line. This means that the slightest variation - a bit being off - can throw us into a different interval, which may correspond to an entirely different sequence. Thus, a small error in the arithmetic code can lead to catastrophic errors in the reconstruction. This is not just a problem with arithmetic codes. All data compression algorithms suffer from this because they entail the removal of redundancy, which is what provides protection against errors. A number of solutions have been proposed for the problem which fall under the general heading of joint source channel coding.

One of the results in Shannon's seminal work [92] was that source coders and channel coders can be designed independently, without reference to each other and with no loss of optimality. The separation theorem has been quite popular as it allowed the design process to be broken down into relatively simpler blocks as shown in Figure 1.1. However, the result is an asymptotic one based on several assumptions. It is assumed that the source coder and channel coder are not constrained in terms of complexity and delay. Furthermore, there is an assumption that the source coder and channel coder operate independently in an optimal fashion. That is, the source coder is able to remove all redundancy from the source, and the channel decoder is able to reproduce the source encoder

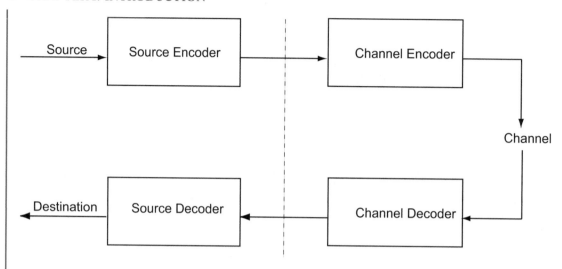

Figure 1.1: Shannon's separate source and channel scheme.

output at the source decoder with negligible distortion. Many of these assumptions do not necessarily hold in reality. For example, the communication systems used between satellite and ground control station need to respond promptly when information is used to control the position of the satellite. In wireless communication, good quality of service requires a limited maximum transmission delay. Krich and Berger [52] showed that when a delay-dependent distortion measure was introduced, the optimal block length associated with minimum distortion was often finite instead of infinite as given in Shannon's theory. It is very difficult to model real world sources and, therefore, practically impossible to design source coders which remove all redundancy from the source. Similarly, in practical situations, it is impossible to guarantee that the channel decoder output will not contain errors which will lead to catastrophic error propagation in the source decoder. In such situations, it may very well make sense to integrate the design of the source and channel coder [86]. Finally, as pointed out by Massey in [62], another missing element in Shannon's separation theorem is the complexity of the separated source and channel coding scheme. Massey [62] and Ancheta [4] showed that joint linear source and channel coding can reduce the overall complexity of the system.

In this book, we explore joint source channel coder designs based on arithmetic coding. We briefly review joint source channel coding algorithms in the following section in order to introduce some terminology and to provide a context for the designs focused on arithmetic codes. The listing is rather cursory and can be skipped on first reading. We will then proceed in the following chapters to a more detailed description of arithmetic coding followed by different methods of joint source channel coding using arithmetic codes.

1.2   JOINT SOURCE AND CHANNEL CODING SCHEMES

In [90], Sayood, Otu and Demir classified the approaches of joint source and channel coding into four categories. We will use the same taxonomic approach. The first class of schemes is denoted as joint source channel coder where the source and channel operations are fully integrated in one coder. Work in this category includes that of Massey [62] and Ancheta [4] on the linear joint source and channel coding scheme, Dunham and Gray [25] on the existence of joint source and noisy channel trellis coding scheme, and Ayanoglu and Gray [5] on the design of the joint source and channel waveform coders.

For a communication system with a fixed bit rate over a noisy channel, we can spend more resources on source coding to achieve a higher SNR if the channel is less noisy. On the other hand, we need to spend more resources on channel coding if the channel is very noisy. For a given channel, we need to make a trade-off in allocating resources between the source coder and channel coder to maximize the overall system performance. This second class of schemes is denoted as concatenated source and channel coder. Early work in this category includes different image coding schemes (Modestino and Daut's two-dimensional differential pulse code modulation (2-D-DPCM) of images [65], Modestino, Daut and Vickers's 2-D discrete cosine transform coding of images [66], and Modestino, Bhaskaran, and Anderson's tree encoding of images [64]) concatenated with convolutional codes, image coding schemes (2-D-DCT) concatenated with Hamming codes by Comstock and Gibson [20], and speech coding schemes concatenated with convolutional codes [67, 77, 32, 33]. Later on, Hochwald and Zeger gave analytical results on choosing the optimal trade-off between source and channel coding for delivering an analog source in [45], and Sherwood and Zeger presented a progressive image coding scheme in [93].

When there are errors present at the input of the source decoder, the error effect depends on the position where the errors occur. For example, the distortion resulting from an error in the scale factor in an MPEG audio stream will be more severe than that of an error in a subband sample. Another category of joint source and channel coding schemes addresses this problem by choosing different error protection levels to protect different parts of the encoded stream according to the error effect. The work in this third class is denoted as unequal error protection (UEP) schemes. This work includes that of Sundberg [97] who first investigated the single bit error effect in a nonlinear PCM system. Shiozaki [94] proposed unequal error protection of PCM signals by self-orthognal convolutional codes. Cox *et al.* [21] analyzed error effects in the subband speech coder and protected the coded stream with matched rate compatible punctured convolutional codes (RCPC). Srinivasan *et al.* [95] chose different protection levels for the quantized subband coefficients according to their importance in the reconstruction to minimize the overall mean-squared distortion of the video. Nazer and Alajali [69] used RCPC codes to protect codebook excited linear prediction (CELP) coded voice streams according to the importance of the bits in the reconstruction to achieve a better output SNR over noisy channel. Ho and Kahn [44] changed the power and modulation of subchannels to provide different protections to the quantized signal to minimize the distortion. Banister, Belzer and Fischer [7] proposed a dynamic programming optimization method to choose UEP schemes for a JPEG2000 coder and turbo codes with a constraint of fixed-size channel packets. Finally, Lan *et al.* [57] presented a fast algorithm for optimal UEP schemes using image or video coders and irregular repeat-accumulate codes. The UEP schemes are very effective and have been adopted by many commercial communication standards such as the global system for mobile communications (GSM) [63] and digital audio broadcasting (DAB) [28].

The fourth and final class of joint source and channel schemes is known as constrained joint source/channel coding where the source coder and/or the decoder are modified to accommodate the noisy channel. We first focus our attention on the decoder. For example, we have work in which source decoders are trying to utilize the remaining redundancy to detect channel errors and to reduce the error effects. This includes the work of Steele, Goodman, and McGonegal [96] who detected transmission errors in a DPCM speech system based on the sample-to-sample differences and replaced the error samples with the output of a smoothing circuit to reduce the error effects, the work of Ngan and Steele [70] who applied a similar idea to PCM and DPCM coded images, the work of Reininger and Gibson [78] who examined the reliability of the decoded 2-D-DCT coefficients of the image and replaced unreliable coefficients with estimates, the work of Sayood and Borkenhagen [86, 87] who performed source sequence estimation to utilize the remaining source output redundancy instead of decoding a symbol at a time, the work of Phamdo and Farvardin [76] who performed the sequence detection of discrete Markov sources over discrete memoryless channels (DMC), and the work of Sayood, Liu, and Gibson [88, 89] who proposed a joint source/channel decoder for binary and nonbinary convolutional codes that incorporate both the source and channel statistics in the path metric.

The idea of sequence estimation is widely used in channel decoding, and its success in source decoding prompted people to apply other channel decoding algorithms to source decoding. Special attention is focused on entropy coders such as Huffman and arithmetic coders since the entropy coder is the last component in a typical source coder [80]. For the list Viterbi decoding algorithm, we have the work of Sayood, Otu and Demir [23, 90] who decoded the Huffman-coded sequences on a symbol-constrained direct graph, the work of Murad and Fuja [68] who combined the source model, Huffman coder and channel coder into one super graph and decoded the source sequence on the trellis derived from the super graph, the work of Park and Miller [71, 72] who decoded the Huffman-coded sequences on a bit-constrained direct graph, and the work of Demiroglu, Hoffman and Sayood [24] who integrated the arithmetic codes with trellis coded modulation and used Euclidean distance and arithmetic coding error detection to prune paths in a list Viterbi decoder. For soft input soft output (SISO) decoding algorithms, we have the work of Wen and Villasenor who present a soft input decoding algorithm in [100] and a SISO decoding algorithm in [101], the work of Bauer and Hagenauer [8, 9] who derived a bit level SISO variable length code (VLC) decoder to concatenate with an inner channel coder to form an iterative decoding scheme, the work of Lakovic, Villasenor, and Wesel [55] who jointly decoded the convolutional codes and Huffman codes in one trellis, the work of Guivarch, Carlach, and Siohan [41], and Jeanne, Carlach and Siohan [48] who derived the bit probability from the symbol probability of Huffman codes and used it in the channel decoder to improve its performance, and the work of Hedayat and Nosratinia [42] who used both SISO VLC decoding and list Viterbi decoding in a concatenated VLC and channel coding scheme.

Another subset of constrained source/channel coding schemes modified the source coders to accommodate the noisy channel. Early work includes that of Kurtenbach and Wintz [53] who designed a scalar quantizer based on the source probability density function and the channel transition matrix, Farvardin and Vaishampayan [30] who further considered the codeword assignments in the scalar quantizer design for the binary symmetric channel (BSC), Vaishampayan and Farvardin [98] who proposed optimization methods to quantize the 2-D-DCT coefficients of an image over the BSC channel, and Chang and Donaldson [18] who derived a closed-form analytical expression for the output signal-to-noise ratio of DPCM systems for noisy channel operation. Most of the recent work in this subset focuses on entropy coders. For Huffman codes, we have the work of Buttigieg [15], and Buttigieg and Farrell [16] who analyzed the distance properties of the Huffman codes and used them as error correcting codes, the work of Laković and Villasenor [54] who proposed an algorithm for construction of reversible variable length codes (RVLCs) with certain free distance properties, the work of Wenisch, Swaszek, and Uht [102] who derived a analytical bound with respect to the Hamming distance of variable length codes, the work of Lamy and Bergot [56] who extended the analytical bound against the sliding distance and free distance, and the work of Wang, Yang and Hanzo [99] who proposed an iterative algorithm to construct reversible variable-length codes and variable-length error-correcting codes.

1.3   JOINT SOURCE AND CHANNEL CODING WITH ARITH-METIC CODES

Boyd et al. [14] proposed a method to enable arithmetic codes to detect errors by reserving probability space for a forbidden symbol that does not belong to the source alphabet in 1997. Later Chou and Ramchandran [19] and Anand, Ramchandran, and Kozintsev [3, 51] analyzed and demonstrated the effectiveness of arithmetic codes with forbidden symbols for error detection. Pettijohn, Hoffman, and Sayood [75, 74] used the reserved probability space in arithmetic coding for error correction and proposed two sequential decoding algorithms, depth first and breadth first, to decode the arithmetic codes against channel errors. Sayir [84] who also investigated arithmetic codes with forbidden symbols and concluded that arithmetic codes are more general codes that can contain block codes and convolutional codes[1]. Elmasry [26, 27] also proposed a different way to incorporate channel coding in the arithmetic coding process.

On the decoding side, we have the work of different decoding algorithms for arithmetic codes with forbidden symbols such as MAP decoding algorithms by Grangetto and Cosman in [34], a soft decoding algorithm based on symbol and bit clock models for arithmetic codes [38] and for a binary quasi-arithmetic decoder with binary and M-ary sources [39, 40] by Guionnet and Guillemot, a modified Bahl, Cocke, Jelinek and Raviv (BCJR) algorithm [6] for the decoding of

---

[1]Jossy Sayir used the term gap instead of a forbidden symbol in [84]. Both terms are used here depending on the context and treated as interchangeable.

arithmetic codes and forming of an iterative scheme using the arithmetic coder with an inner channel coder by Grangetto, Scanavino, and Olmo [36], an iterative decoding scheme using a soft arithmetic decoder and a hard arithmetic decoder proposed by Xu, Hoffman, and Sayood [105], and a trellis decoding scheme where a finite state machine model is used to interpret arithmetic codes with forbidden symbols as variable-input-length, variable-output-length trellis codes by Bi, Hoffman, and Sayood [12]. The hard decision error correction scheme proposed in [105] is further explored by Zhang, Wang and Zhou in [107]. This method is also being applied with trellis decoding to form an iterative decoding scheme in [59].

Because of their superior performance, arithmetic codes with forbidden symbol are proposed for real world applications. Grangetto, Magli and Olmo applied an arithmetic code with a forbidden symbol in the JPEG 2000 frame work in [35] and achieved better performance. Logashanmugam, Sreejaa and Ramachndran proposed using arithmetic coding with a forbidden symbol for H.264 Video coders in [43].

Quasi-Arithmetic (QA) codes have been presented in [39, 40] in several ways. In [39, 40] the error protection is provided in two ways: using the arithmetic codes with a marker symbol or by serially concatenating inner convolutional codes with outer arithmetic codes. While arithmetic codes with markers do provide resynchronization and low bit error rates, frame error rates remain high. Consequently, for low frame error rates the concatenated inner convolutional codes and outer arithmetic codes from [39, 40] would be preferred. These codes are subsequently iteratively decoded. The decoding trellis in [39, 40] is derived from the standard arithmetic decoder and the state variables consist of two intervals, the updated interval from current bit input and the interval from the last decoded symbol. Also, in [39, 40], the number of states in the decoding trellis grows quadratically with the packet length since the number of decoded symbols is used as part of the state variable description in order for the trellis paths to remain separated. For example, in [39, 40], there are an average of 1193 states when $N = 4$ ([0, N] is the integer interval) and 4736 states when $N = 8$. A recent paper [60] by Malinowski, Jegou and Guillemot addresses this issue by using the number of symbols modulo a length constraint parameter, T, with the result that some paths are indistinguishable although the overall impact on error rate is low.

On the distance property of arithmetic codes with forbidden symbols, a similar work [10] is independently developed in parallel with the work presented here. In their work, Ben-Jama et al. [10] reduced the three-dimensional trellis used in this work into a two-dimensional trellis by adding branch transitions to eliminate state transitions without any outputs.

We briefly review arithmetic codes in the following chapter. In Chapter 3, Arithmetic codes with forbidden symbols are introduced and a finite state machine interpretation of Arithmetic codes is presented. We show how using list Viterbi decoding on the trellis representation of the state machine can provide error correction. We also extend this state machine interpretation to arithmetic coding with forbidden symbols for Markov sources. In Chapter 4, we analyze the distance properties of arithmetic codes with forbidden symbols and use the partial distance spectrum as a tool for code selection. Arithmetic codes with forbidden symbols with memory are also proposed in this chapter and new constructed codes are simulated and compared with state-of-the-art turbo codes. Finally, we present some conclusions in Chapter 5.

CHAPTER 2

Arithmetic Codes

The idea of arithmetic coding can be traced back to Shannon's 1948 original paper [92], which mentioned an approach using the cumulative distribution function to generate a binary code for a sequence (a technique now known as Shannon-Fano coding). Elias came up with the idea of recursive implementation, which was first mentioned in [2] and appeared in [49].

Before we get into the details of the arithmetic coding process, we need to establish some notation. Let A be a discrete source with a finite alphabet $\{a_1, a_2, \cdots, a_M\}$. The source symbol can be mapped into the integers $\{1, 2, \cdots, M\}$ by $X(a_i) = i$, $a_i \in A, i \in Z$, where X is a random variable. Let $S = \{s_1, s_2, \cdots s_N\}$ be a sequence of N random symbols generated by source A. We assume the source symbols are independent and identically distributed with probability function $P(i) = P(a_i)$ and cumulative probability function $F_X(i) = \sum_{k=1}^{i} P(a_k)$.

The basic idea of arithmetic coding is to map the input source sequences into disjoint subintervals of a given initial interval (typically $[0, 1)$ is chosen). This can be done by a recursive partitioning of the interval driven by the source sequence. We use $[l, u)$ to denote the lower and the upper limits of the interval, and use $l^{(n)}, u^{(n)}$ to denote, respectively, the lower limit and upper limit after encoding the nth symbol. Another popular way to represent the interval uses the interval base and interval length pair. There is no mathematical difference between these two notations. The recursive partitioning process can be described as follows [85],

$$l^{(0)} = 0, \tag{2.1}$$
$$u^{(0)} = 1, \tag{2.2}$$

and

$$l^{(n)} = l^{(n-1)} + (u^{(n-1)} - l^{(n-1)})F_X(X(s_n) - 1), \tag{2.3}$$
$$u^{(n)} = l^{(n-1)} + (u^{(n-1)} - l^{(n-1)})F_X(X(s_n)). \tag{2.4}$$

After the interval for a specific source sequence is known, any value in this interval can be used to represent this interval. We would like to choose a value in the interval that can be represented by the smallest number of bits. The minimum number of bits required for representing the interval is $\lceil -log_2(\Delta l) \rceil$, where Δl is the interval length and $\lceil x \rceil$ represents the smallest integer greater than or equal to x.

2.1 ENCODING AND DECODING PROCESSES

Following the recursive equations 2.3 and 2.4, we can map an input sequence into an interval. For example, consider the source given in Table 2.1.

Table 2.1: Source with three symbols.

Symbol	Probability	Cumulative probability
a_1	0.4	0.4
a_2	0.35	0.75
a_3	0.25	1.0

Example 2.1   Suppose we want to encode an input sequence $a_1a_3a_2a_1$. We first need to initialize $l^{(0)}$ to 0, $u^{(0)}$ to 1. Then following the update equation 2.3 and 2.4 to encode a_1, we have updated interval [0, 0.4). After encode the second symbol a_3, the interval becomes [0.3, 0.4). Encoding the third element of the sequence a_2 results in interval [0.34, 0.375). With the last symbol a_1, the arithmetic coding process generates the final interval [0.34, 0.354). Any point in this interval can be used as a tag to represent the input sequence. If we choose the middle point of the interval as the tag, then

$$T_X(a_1a_3a_2a_1) = \frac{0.34 + 0.354}{2} = 0.347.$$

Decoding is a recursive process: initialize the interval, partition the interval according source cumulative probability distribution and decoding out according to where tag resides, update the interval, then repeat above procedure.

In the above example, we first set the initial interval with $l^{(0)} = 0$ and $u^{(0)} = 1$. Partition the interval according to the source cumulative probability distribution. Since the tag 0.347 resides in the interval [0, 0.4), we decode first the symbol as a_1. Partition the interval [0, 0.4) again, and the tag falls into the interval [0.3, 0.4), this generates a second output a_3. Repeating the above process, we can decode all the symbols.

This algorithm has two drawbacks. First, in a fixed precision implementation, the above algorithm results in a smaller and smaller interval as we continuously partition the interval according to the input symbols. The interval needs to be renormalized or rescaled whenever it becomes too small. The second problem is a long encoding delay. The binary outputs are generated only after the last symbol is encoded. Over the years different encoding and decoding schemes are developed to deal with the fixed precision implementation challenges by Pasco [73], Rissanen [79], Rubin [82], Martin [61], Rissanen and Langdon [81], Guzzo [37], and Witten, Neal, and Cleary [103]. There are many different ways to renormalize and generate incremental output. The following is the implementation taken from [85], which was proposed by Witten, Neal and Cleary in [103]. The interval is rescaled (mapped) as described while *Scale3* is a special register that records the number of E_3 mappings that have occurred.

- E_1: The interval is entirely confined to the lower half of the unit interval, i.e, $0 \leq u < 0.5$. Perform an E_1 mapping on the lower and upper limits of the current interval and output 0, where $E_1(x) = 2x, \ x = l, u.$

- E_2: The interval is entirely confined to the upper half of the unit interval, i.e, $0.5 \leq l < 1.0$. Perform an E_2 mapping on the lower and upper limits of the current interval and output 1, where $E_2(x) = 2(x - 0.5)$, $x = l, u$.

- E_3: The interval straddles the midpoint of the unit interval and $0.25 \leq l < 0.5 \leq u < 0.75$. Perform an E_3 mapping on the lower and upper limit of the current interval and increase $Scale3$ by 1, where $E_3(x) = 2(x - 0.25)$, $x = l, u$.

The encoder is initialized with $Scale3$ set equal to 0 and the lower and upper limits set to their extreme values. The encoding process begins with the updates given in Equations 2.3 and 2.4. The interval is rescaled whenever one of the above three cases occur. Whenever an E_1 or an E_2 mapping occurs, one bit indicates the mapping followed by $Scale3$ bits, which are the complement of the bit indicating the E_1 or E_2 mapping. $Scale3$ is then set to 0. For example, suppose $Scale3$ is 5, if an E_1 mapping happens, the encoder will first output a 0 for the E_1 mapping followed by 11111, and $Scale3$ is then set to 0; if an E_2 mapping happens, the encoder will output a 1 for the E_2 mapping followed by 00000, and $Scale3$ is then set to 0.

2.2   INTEGER IMPLEMENTATION OF ENCODING AND DE-CODING WITH RENORMALIZATION

Most practical implementations use integer arithmetic and generate the binary code in the process. We use a set of integer registers to represent the probabilities and the cumulative probabilities of the source. For example, if an integer $Total\_count = 20$ is used to represent the maximum occurring frequency of the source, we can use the following integers in Table 2.2 to represent the probabilities and cumulative probabilities in Table 2.1. The source symbol probabilities and cumulative probabilities can be computed as

$$P(a_i) = \frac{Count(a_i)}{Total\_count}$$
$$F_X(X(a_i)) = \frac{Cum\_Count(a_i)}{Total\_count}.$$

Table 2.2: Source with three symbols.

Symbol	Count	Cum_Count
a_1	8	8
a_2	7	15
a_3	5	20

We use the following integer arithmetic update equations to replace equations 2.1-2.4,

$$l^{(0)} = 0, \tag{2.5}$$
$$u^{(0)} = Top\_value, \tag{2.6}$$

and

$$l^{(n)} = l^{(n-1)} + \left\lfloor \frac{(u^{(n-1)} - l^{(n-1)} + 1) \times Cum\_Count(X(x_n) - 1)}{Total\_Count} \right\rfloor, \tag{2.7}$$

$$u^{(n)} = l^{(n-1)} + \left\lfloor \frac{(u^{(n-1)} - l^{(n-1)} + 1) \times Cum\_Count(X(x_n))}{Total\_Count} \right\rfloor - 1 \tag{2.8}$$

where the Top_value is the maximum integer used to represent the upper limit of the interval (corresponding to 1 in the unit interval) and $\lfloor x \rfloor$ represents the greatest integer that less or equal to x. In practice, we would like to choose $Top\_value = 2^m - 1$ since this makes both the rescaling checking and rescaling operations much easier. The intervals need to be rescaled if the lower and upper limits of the interval have the following binary forms

$$E_1 \ mapping : \begin{cases} l^{(n)} = 0XXX \cdots \\ u^{(n)} = 0XXX \cdots \end{cases}, \tag{2.9}$$

$$E_2 \ mapping : \begin{cases} l^{(n)} = 1XXX \cdots \\ u^{(n)} = 1XXX \cdots \end{cases}, \tag{2.10}$$

and

$$E_3 \ mapping : \begin{cases} l^{(n)} = 01XXX \cdots \\ u^{(n)} = 10XXX \cdots \end{cases}. \tag{2.11}$$

The source sequences can be encoded and decoded with the same procedure except for using the integer arithmetic of equations 2.5-2.8. We repeat the arithmetic encoding of input sequence $a_1 a_3 a_2 a_1$ using the probability counts in Table 2.2 with integer arithmetic.

2.2.1   ENCODING WITH INTEGER ARITHMETIC

Example 2.2   First, we reset $Scale3$ and the initial coding interval by choosing $Top\_value = 127$ with $m = 7$,

$$\begin{aligned} Scale3 &= 0 \\ l^{(0)} &= 0 \\ u^{(0)} &= 127. \end{aligned}$$

To update the interval for input a_1 we compute

$$l^{(1)} = 0 + \left\lfloor \frac{(127 - 0 + 1) \times 0}{20} \right\rfloor = 0 = (0000000)_2$$

$$u^{(1)} = 0 + \left\lfloor \frac{(127 - 0 + 1) \times 8}{20} \right\rfloor - 1 = 50 = (0110010)_2.$$

Since the interval satisfies the E_1 rescaling condition (2.9), the encoder performs an E_1 mapping by left shifting one bit and outputting a 0. The LSB of the upper limit is set to 1 and the LSB of the lower limit is set to 0. The interval is changed to

$$
\begin{aligned}
l'^{(1)} &= (0000000)_2 = 0 \\
u'^{(1)} &= (1100101)_2 = 101.
\end{aligned}
$$

To update the interval for the second symbol a_3 we compute

$$
\begin{aligned}
l^{(2)} &= 0 + \left\lfloor \frac{(101 - 0 + 1) \times 15}{20} \right\rfloor = 76 = (1001100)_2 \\
u^{(2)} &= 0 + \left\lfloor \frac{(101 - 0 + 1) \times 20}{20} \right\rfloor - 1 = 101 = (1100101)_2.
\end{aligned}
$$

Since the interval satisfies the E_2 rescaling condition (2.10), the encoder performs an E_2 mapping by left shifting one bit and outputting a 1. The LSB of the upper limit is again set to 1, and the LSB of the lower limit is set to 0. The interval is changed to

$$
\begin{aligned}
l'^{(2)} &= (0011000)_2 = 24 \\
u'^{(2)} &= (1001011)_2 = 75.
\end{aligned}
$$

Updating the interval with input a_2, we have

$$
\begin{aligned}
l^{(3)} &= 24 + \left\lfloor \frac{(75 - 0 + 1) \times 8}{20} \right\rfloor = 54 = (0110110)_2 \\
u^{(3)} &= 24 + \left\lfloor \frac{(75 - 0 + 1) \times 15}{20} \right\rfloor - 1 = 80 = (1010000)_2.
\end{aligned}
$$

Since the interval satisfies the E_3 rescaling condition (2.11), the encoder performs an E_3 mapping by left shifting one bit and complementing the second MSB. The LSB of the upper limit is set to 1, and the LSB of the lower limit is set to 0. The interval and $Scale3$ are changed to

$$
\begin{aligned}
Scale3 &= Scale3 + 1 = 1 \\
l'^{(3)} &= (0101100)_2 = 44 \\
u'^{(3)} &= (1100001)_2 = 97.
\end{aligned}
$$

Encoding the last symbol a_1, we have

$$
\begin{aligned}
l^{(4)} &= 24 + \left\lfloor \frac{(97 - 44 + 1) \times 0}{20} \right\rfloor = 24 = (0011000)_2 \\
u^{(4)} &= 24 + \left\lfloor \frac{(97 - 44 + 1) \times 8}{20} \right\rfloor - 1 = 44 = (0101100)_2.
\end{aligned}
$$

This interval satisfies the E_1 condition, so the encoder outputs 0 followed by a 1, resets $Scale3$ to zero and updates the interval,

$$
\begin{aligned}
Scale3 &= 0 \\
l''^{(4)} &= (0110000)_2 = 48 \\
u''^{(4)} &= (1011001)_2 = 81.
\end{aligned}
$$

The tag will be

$$
T_X(a_1a_3a_2a_1) = 48 + \left\lfloor \frac{81 - 48 + 1}{2} \right\rfloor = 65 = (1000001)_2.
$$

Combining with the previous encoder output $(0101)_2$, we have the total output tag of $(01011000001)_2$.

2.2.2   DECODING WITH INTEGER ARITHMETIC

Example 2.3   We first set the initial interval to $l^{(0)} = 0, u^{(0)} = 127$ and we set $Scale3$ to 0. When the first symbol s_1 is encoded as follows,

$$
\begin{aligned}
l^{(1)} &= 0 + \left\lfloor \frac{(127 - 0 + 1) \times F_X(X(x_1) - 1)}{20} \right\rfloor = \left\lfloor \frac{32 \times F_X(X(x_1) - 1)}{5} \right\rfloor \\
u^{(1)} &= 0 + \left\lfloor \frac{(127 - 0 + 1) \times F_X(X(x_1) - 1)}{20} \right\rfloor - 1 = \left\lfloor \frac{32 \times F_X(X(x_1))}{5} - 1 \right\rfloor.
\end{aligned}
$$

The possible intervals becomes $[0, 50], [51, 95]$ and $[96, 127]$ for $s_1 = a_1, a_2$ and a_3, respectively. Since $m = 7$, the first seven bits of the tag are $(0101100)_2 = 44$, which lies in the interval $[0, 50]$. Hence, the first symbol is decoded as a_1. Performing an E_1 mapping on the interval and the tag gives

$$
\begin{aligned}
l'^{(1)} &= (0000000)_2 = 0 \\
u'^{(1)} &= (1100101)_2 = 101 \\
T_X &= (1011000)_2 = 88,
\end{aligned}
$$

where the tag shifts one bit out to the left and sets the LSB of the tag to the next tag sequence bit. When we update the interval with the second symbol s_2, the possible intervals become

$$
\begin{aligned}
l^{(2)} &= 0 + \left\lfloor \frac{(101 - 0 + 1) \times F_X(X(s_2) - 1)}{20} \right\rfloor = \left\lfloor \frac{51 \times F_X(X(s_2) - 1)}{10} \right\rfloor \\
u^{(2)} &= 0 + \left\lfloor \frac{(127 - 0 + 1) \times F_X(X(s_2) - 1)}{20} \right\rfloor - 1 = \left\lfloor \frac{51 \times F_X(X(s_2))}{10} \right\rfloor - 1
\end{aligned}
$$

[0, 40], [41, 75], and [76, 101] for $s_2 = a_1, a_2$, and a_3, respectively. The second symbol is decoded as a_3 since the current tag value of 88 is contained in that interval. Performing an E_2 mapping on the interval and the tag gives

$$
\begin{aligned}
l'^{(2)} &= (0011000)_2 = 24 \\
u'^{(2)} &= (1001011)_2 = 75 \\
T_X &= (0110000)_2 = 48.
\end{aligned}
$$

Repeating the above process, we have encoded the third symbol s_3 as follows,

$$
\begin{aligned}
l^{(3)} &= 24 + \left\lfloor \frac{(75 - 24 + 1) \times F_X(X(s_3) - 1)}{20} \right\rfloor = 24 + \left\lfloor \frac{13 \times F_X(X(s_3) - 1)}{5} \right\rfloor \\
u^{(3)} &= 24 + \left\lfloor \frac{(75 - 24 + 1) \times F_X(X(s_3) - 1)}{20} \right\rfloor - 1 = 23 + \left\lfloor \frac{13 \times F_X(X(s_3) - 1)}{5} \right\rfloor.
\end{aligned}
$$

So the possible intervals are [24, 43], [44, 62], and [63, 75] for $s_3 = a_1, a_2$, and a_3, respectively. The current tag value of 48 lies in the second interval, so the third symbol is decoded as $s_3 = a_2$. Since the interval satisfies the E_3 mapping condition, with E_3 mapping, we increase $Scale3$ and update the interval

$$
\begin{aligned}
Scale3 &= Scale3 + 1 = 1 \\
l'^{(3)} &= (0101100)_2 = 44 \\
u'^{(3)} &= (1100001)_2 = 97 \\
T_X &= (0110000) = 48.
\end{aligned}
$$

Since the E_3 mapping does not generate any output, the tag is kept the same. With the same procedure, we can encode the last symbol s_4 of the input sequence,

$$
l^{(4)} = 44 + \left\lfloor \frac{(97 - 44 + 1) \times F_X(X(s_4) - 1)}{20} \right\rfloor = 44 + \left\lfloor \frac{27 \times F_X(X(s_4) - 1)}{10} \right\rfloor
$$

$$
u^{(4)} = 44 + \left\lfloor \frac{(97 - 24 + 1) \times F_X(X(s_4))}{20} \right\rfloor = 43 + \left\lfloor \frac{27 \times F_X(X(s_4))}{10} \right\rfloor
$$

for all the possible input symbols, $s_4 = a_1, a_2$, and a_3, the resulting intervals are [44, 64], [65, 83], and [84, 97], respectively. The tag value of 48 resides on the interval associated with symbol a_1. Hence, we have the decoded sequence $a_1 a_3 a_2 a_1$.

2.2.3   OVERFLOW AND UNDERFLOW PROBLEMS

2.2.3.1 Underflow

When the interval is updated using equations 2.7 and 2.8, we need to avoid underflow [103] where we obtain interval limits that violate $l^{(n)} < u^{(n)}$. The negative interval will cause improper encoding and decoding. Given the value of $Total\_Count$, which is used to represent the occurring frequency of the input symbols, we can choose the Top_value (or, alternately the number of tag bits (m) to avoid the underflow problem. We can rewrite equation 2.8 and have

$$
\begin{aligned}
u^{(n)} &= l^{(n-1)} + \left\lfloor \frac{(u^{(n-1)} - l^{(n-1)} + 1) \times Cum\_Count(X(s_n))}{Total\_Count} \right\rfloor - 1 \\
&\geq l^{(n-1)} + \left\lfloor \frac{(u^{(n-1)} - l^{(n-1)} + 1) \times Cum\_Count(X(s_n) - 1)}{Total\_Count} \right\rfloor \\
&\quad -1 + \left\lfloor \frac{(u^{(n-1)} - l^{(n-1)} + 1) \times Count(X(s_n))}{Total\_Count} \right\rfloor \\
&= l^{(n)} + \left\lfloor \frac{(u^{(n-1)} - l^{(n-1)} + 1) \times Count(X(s_n))}{Total\_Count} \right\rfloor - 1.
\end{aligned}
$$

To avoid the underflow problem, it is sufficient to let

$$
\left\lfloor \frac{(u^{(n-1)} - l^{(n-1)} + 1) \times Count(X(x_n))}{Total\_Count} \right\rfloor - 1 \geq 0.
$$

After removing the $\lfloor \ \rfloor$ function, it becomes

$$
(u^{(n-1)} - l^{(n-1)} + 1) \times Count(X(x_n)) \geq Total\_Count.
$$

Let Min_Count denote the smallest count among the input symbols, then it it is sufficient to bound the previous interval length $u^{(n-1)} - l^{(n-1)} + 1$ to avoid underflow

$$
(u^{(n-1)} - l^{(n-1)} + 1) \geq \frac{Total\_Count}{Min\_Count}. \tag{2.12}
$$

If $l^{(n)}$ and $u^{(n)}$ are in the forms given in 2.9-2.11, they will invoke the rescaling mappings. So the interval limits will have one of the following forms before updating using the next symbol,

$$
\begin{cases}
l^{(n)} &= 0XXXX \cdots \\
u^{(n)} &= 11XXX \cdots
\end{cases} \tag{2.13}
$$

$$
\begin{cases}
l^{(n)} &= 00XXX \cdots \\
u^{(n)} &= 1XXXX \cdots
\end{cases} \tag{2.14}
$$

The smallest length of an interval that can be reached under the above scheme will be

$$
\frac{Top\_value + 1}{4} + 1.
$$

We need to make sure that the smallest interval satisfies equation 2.12. This leads to

$$Top\_value \geq 4 \times (\frac{Total\_Count}{Min\_Count} - 1) - 1. \qquad (2.15)$$

We can also fix Top_value and put a constraint on $Total\_Count$ as

$$Total\_Count \leq Min\_Count \times (\frac{(Top\_value + 1)}{4} + 1). \qquad (2.16)$$

2.2.3.2 Overflow

In order to avoid multiplication overflow problem in equation 2.7 and 2.8, the integer arithmetic needs to be able to handle numbers within the following numerical range

$$0 \leq x \leq Total\_Count \times (Top\_value + 1).$$

2.3   OPTIMALITY OF ARITHMETIC CODING

As is mentioned previously, the minimum number of binary digits required to represent the coding interval is $\lceil -log_2(\Delta l) \rceil$. Here we will show that there exists a number within an arbitrary interval $[x, x + \Delta l)$ that can be expressed by $\lceil -log_2(\Delta l) \rceil$ binary bits. Let $k = \lceil -log_2(\Delta l) \rceil$, and we express x and $x + \Delta l$ in binary forms

$$\begin{cases} x & = 0.\; \underbrace{x_1 x_2 \cdots x_k}_{k} XXX \cdots \\ x + \Delta l & = 0.\; \underbrace{y_1 y_2 \cdots y_k}_{k} XXX \cdots \end{cases}.$$

Since

$$\begin{aligned} k & = \lceil -log_2(\Delta l) \rceil \\ & \geq -log_2(\Delta l), \end{aligned}$$

we have

$$\Delta l \geq \frac{1}{2^k}.$$

This results in at least one bit difference on the first k bits of the x and $x + \Delta l$. Hence, there exists a number z that has the form of $0.\; \underbrace{z_1 z_2 \cdots z_k}_{k} 000 \cdots$ and $x \leq z \leq x + \Delta l$. It can be seen that we need some extra operations to find the number z within the final interval. In practice, we would like to truncate the $\lceil -log_2(\Delta l) \rceil + 1$ bit of the binary representation of the middle point of the final interval, i.e, the tag, since it is simple, and we only have one bit overhead for the entire coded sequence.

2.3.1   ARITHMETIC CODES ARE PREFIX CODES

A code is a prefix code when no codeword is a prefix of another codeword. Unlike Huffman coding, which generates a unique codeword for each input symbol, arithmetic coding produces a unique tag for each input sequence. Let $\mathbf{x}$ and $\mathbf{y}$ be two distinct sequences of source A; their probabilities are $P(\mathbf{x})$ and $P(\mathbf{y})$, and their arithmetic coding intervals are $[F_X(X(\mathbf{x}) - 1), \; F_X(X(\mathbf{x}) - 1) + P(\mathbf{x}))$ and $[F_X(X(\mathbf{y}) - 1), \; F_X(X(\mathbf{y}) - 1) + P(\mathbf{y}))$, respectively. These two intervals are disjoint. The encoded outputs of $\mathbf{x}$ are the truncation of the $k(\mathbf{x}) = \lceil -log_2(P(\mathbf{x})) \rceil + 1$ bits of the binary representation of the tag $T_X(\mathbf{x})$. If $\lfloor T_X(\mathbf{x}) \rfloor_{k(\mathbf{x})}$ is used to represent the truncated tag value, for prefix condition it is sufficient to show that interval $[\lfloor T_X(\mathbf{x}) \rfloor_{k(\mathbf{x})}, \; \lfloor T_X(\mathbf{x}) \rfloor_{k(\mathbf{x})} + \frac{1}{2^{k(\mathbf{x})}})$ lies inside of interval $[F_X(X(\mathbf{x}) - 1), \; F_X(X(\mathbf{x}) - 1) + P(\mathbf{x}))$ [85], i.e., the numbers of form $\lfloor T_X(\mathbf{x}) \rfloor_{k(\mathbf{x})} XXX \cdots$ are totally contained in interval $[F_X(X(\mathbf{x}) - 1), \; F_X(X(\mathbf{x}) - 1) + P(\mathbf{x}))$, and $\lfloor T_X(\mathbf{x}) \rfloor_{k(\mathbf{x})}$ can not be a prefix of any number in $[F_X(X(\mathbf{y}) - 1), \; F_X(X(\mathbf{y}) - 1) + P(\mathbf{y}))$. Hence, we need to show $F_X(X(\mathbf{x}) - 1) < \lfloor T_X(\mathbf{x}) \rfloor_{k(\mathbf{x})} < \lfloor T_X(\mathbf{x}) \rfloor_{k(\mathbf{x})} + \frac{1}{2^{k(\mathbf{x})}} \leq F_X(X(\mathbf{x}) - 1) + P(\mathbf{x})$. We first show $\lfloor T_X(\mathbf{x}) \rfloor_{k(\mathbf{x})} > F_X(X(\mathbf{x}) - 1)$. Since

$$0 \leq T_X(\mathbf{x}) - \lfloor T_X(\mathbf{x}) \rfloor_{k(\mathbf{x})} < \frac{1}{2^{k(\mathbf{x})}}$$

and

$$\begin{aligned} \frac{1}{2^{k(\mathbf{x})}} &= \frac{1}{2^{-\lceil log_2(P(\mathbf{x})) \rceil + 1}} \\ &\leq 2^{log_2(P(\mathbf{x})) - 1} \\ &= \frac{P(\mathbf{x})}{2} \end{aligned}$$

we have

$$0 \leq T_X(\mathbf{x}) - \lfloor T_X(\mathbf{x}) \rfloor_{k(\mathbf{x})} < \frac{P(\mathbf{x})}{2}. \tag{2.17}$$

Recall that the tag is defined as,

$$\begin{aligned} T_X(\mathbf{x}) &= \frac{F_X(X(\mathbf{x}) - 1) + (F_X(X(\mathbf{x}) - 1) + P(\mathbf{x}))}{2} \\ &= F_X(X(\mathbf{x}) - 1) + \frac{P(\mathbf{x})}{2} \end{aligned}$$

Taking $T_X(\mathbf{x})$ into equation 2.17, we have

$$\lfloor T_X(\mathbf{x}) \rfloor_{k(\mathbf{x})} > F_X(X(\mathbf{x}) - 1).$$

On the other hand,

$$\begin{aligned} F_X(X(\mathbf{x}) - 1) + P(\mathbf{x})) &= (F_X(X(\mathbf{x}) - 1) + \frac{P(\mathbf{x})}{2}) + \frac{P(\mathbf{x})}{2} \\ &= T_X(\mathbf{x}) + \frac{P(\mathbf{x})}{2} \\ &\geq \lfloor T_X(\mathbf{x}) \rfloor_{k(\mathbf{x})} + \frac{1}{2^{k(\mathbf{x})}}. \end{aligned}$$

Hence, arithmetic codes are prefix codes and unique decodable.

2.3.2   EFFICIENCY

The average code length, $\overline{K}$, of an arithmetic code for a sequence of length M is defined as

$$
\begin{aligned}
\overline{K} &= \sum P(\mathbf{x})k(\mathbf{x}) \\
&= \sum P(\mathbf{x})\left[\left\lceil \log_2 \frac{1}{P(\mathbf{x})} \right\rceil + 1\right] \\
&= \sum P(\mathbf{x})\left[\log_2 \frac{1}{P(\mathbf{x})} + 2\right] \\
&= -\sum P(\mathbf{x}) \log_2 \frac{1}{P(\mathbf{x})} + 2\sum P(\mathbf{x}) \\
&= H(A^M) + 2.
\end{aligned}
$$

Since the average length can not be smaller than the entropy, the average length is bounded by

$$
H(A^M) \leq \overline{K} \leq H(A^M) + 2.
$$

The average length per source symbol, $\overline{k}$, is bounded by

$$
\frac{H(A^M)}{M} \leq \overline{k} \leq \frac{H(A^M) + 2}{M}.
$$

If the source is independent and identically distributed, we have

$$
H(A^M) = MH(A).
$$

$$
H(A) \leq \overline{k} \leq H(A) + \frac{2}{M},
$$

and it follows that

$$
\lim_{m \to \infty} \overline{k} = H(A).
$$

Hence, the arithmetic code is an entropy encoder.

2.3.3   EFFICIENCY OF THE INTEGER IMPLEMENTATION

There are two major factors that affect the efficiency of integer implementations of the arithmetic coding, which are

1. The use of integer arithmetic instead of infinite precision arithmetic and

2. The limitation on *Total_Count* and *Top_value*.

As pointed by Said [83], integer arithmetic can be viewed as the approximation of the exact infinite precision arithmetic, and we can convert this approximation problem of integer arithmetic into an approximation in our modeling the source probabilities. For example, when we encode the first symbol a_1 in section 2.1, the exact interval is $[0, 0.4)$, while with integer implementation, the interval becomes $[0, 50]$. Converting the integer intervals into unit intervals with an open upper limit, we have its corresponding interval $[0, 51/128)$.[1] This is equivalent to changing the interval calculation of

$$
\begin{aligned}
u^{(1)} &= l^{(0)} + (u^{(0)} - l^{(0)}) \times P(a_1) \\
&= 0 + 1.0 \times 0.4
\end{aligned}
$$

into interval calculation of

$$
\begin{aligned}
u'^{(1)} &= l^{(0)} + (u^{(0)} - l^{(0)}) \times P'(a_1) \\
&= 0 + 1.0 \times \frac{51}{128} \\
&= 0 + 1.0 \times 0.3984375
\end{aligned}
$$

Any mismatch between the source and model used for encoding will degrade the compression efficiency. Although the above treatment results in different $P'(a_i)$'s for each round of encoding, we can bound this difference by

$$
1 - \epsilon \leq \frac{P(a_i)}{P'(a_i)} \leq 1 + \epsilon, \ \forall i,
$$

where ϵ is a small number. The average loss is

$$
\begin{aligned}
\Delta &= -\sum_i (P(a_i) \log_2 P'(a_i)) + \sum_i (P(a_i) \log_2 P(a_i)) \\
&= \sum_i \log_2 \frac{P(a_i)}{P'(a_i)} \\
&\leq \sum_i \log_2 (1 + \epsilon).
\end{aligned}
$$

The bound on the inaccuracy of the probability is closely related to the precision used in the integer arithmetic. In practice, the loss of efficiency with integer implementation is very small since 32-bit fixed precision is supported by most of processors, and the resulting loss is less than 1%. The finite precision and underflow problem result in a limitation on the maximum value of $Total\_Count$. This puts a constraint on how accurate we can represent a symbol probability. The problem is the same as the influence of the approximation of integer arithmetic except we have a fix $P'(a_i)$ all the time.

---

[1] In the integer implementation, the LSB bit in the upper limit is always filled with 1 when performing left shifting. So converting the upper limit in the infinite precision, we have infinite number of 1s pending after the first m bits of binary form of the upper limit. For example, here we have $50 = (\underbrace{0110010}_{m}.111\cdots)_2 = 51$.

CHAPTER 3

Arithmetic Codes with Forbidden Symbols

Arithmetic coding is very sensitive to errors. If an error has occurred in the received binary tag sequence, it will cause error propagation in the decoded sequence. For example, if the first bit of the tag $(01011000001)_2$ in section 2.2 is flipped, the tag is changed into $(11011000001)_2$. This bit sequence will be decoded as $a_3a_2a_3a_3$ instead of $a_1a_3a_2a_1$. Boyd et al. [14] proposed to reduce the current coding interval by a *reduction factor* whenever the arithmetic encoder renormalizes the code interval. The error will be detected when tag resides outside of the reduced code interval. Chou and Ramchandran [19] assigned the unused code intervals to forbidden symbols and performed arithmetic coding based on expanded source alphabets. A transmission error is detected when a forbidden symbol is decoded. Figure 3.1 demonstrates the interval partitions with an introduced forbidden symbol in the source alphabet where the reserved probability is ϵ.

3.1 ERROR DETECTION AND CORRECTION USING ARITHMETIC CODES

When a probability space of ϵ is reserved for forbidden symbols, we need to adjust the source probabilities in the source model.

3.1.1 RESERVED PROBABILITY SPACE AND CODE RATE

One common method is employed to shrink the source probabilities by a factor of $1 - \epsilon$, i.e, $P'(a_i) = P(a_i) \times (1 - \epsilon)$. The reserved probability space results in codeword space expansion, adding redundancy. The average code length for an iid source using this new probability model will be

$$
\begin{aligned}
\overline{k'} &= -\sum_i P(a_i) \log_2 P'(a_i) \\
&= -\sum_i P(a_i) \log_2 (P(a_i) \times (1 - \epsilon)) \\
&= -\sum_i P(a_i) \log_2 (P(a_i) - \log_2 (1 - \epsilon)) \\
&= H(A) - \log_2 (1 - \epsilon)
\end{aligned}
$$

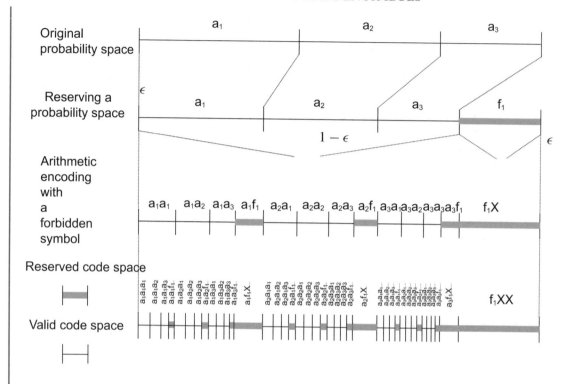

Figure 3.1: Encoding with a forbidden symbol with probability ϵ.

The introduced redundancy is $-\log_2(1-\epsilon)$ bits per input symbol. The corresponding channel coding rate R is given by,

$$R = \frac{H(A)}{H(A) - \log_2(1-\epsilon)}.$$

It can be seen that the code rate depends on both the source entropy $H(A)$ and the reserved probability ϵ in the source space. If the code rate R is fixed, then probability space ϵ that need to be reserved is

$$\epsilon = 1 - (\frac{1}{2})^{(\frac{1}{R}-1)H(A)}. \tag{3.1}$$

One drawback of the source probability shrinking method described above is that the arithmetic encoder generates sequences with unequal probabilities. Using this encoding scheme, the average encoded output length for each particular symbol a_i will be $-\log_2 P(a_i) - \log_2(1-\epsilon)$, i.e, the extra bits required for each source symbol a_i is set at $-\log_2(1-\epsilon)$ bits no matter what the source probability of a_i is. Hence, we can have a situation in which two sequences have the same output length but quite different sequence probabilities. For example, suppose we have an iid source with the probability distribution listed in Table 3.1.

Table 3.1: Source with four symbols.

Symbol	Probability
a_1	0.5
a_2	0.25
a_3	0.125
a_4	0.125

We reserve a probability of $\epsilon = 0.5$ in the source probability space and encode the source sequence with probabilities given in Table 3.2. The arithmetic coder will generate the following outputs, $a_1 \rightarrow 10$, $a_2 \rightarrow 110$, $a_3 \rightarrow 1110$, and $a_4 \rightarrow 1111$. For input sequence $a_1 a_1$, the encoder will output 1010 which has the same length with an output 1111, but the probability of input $a_1 a_1$ is 0.25 while the probability of input a_4 is 0.125. We can avoid such situation using a different method

Table 3.2: The probability table for encoding a source with four symbols after inserting a forbidden symbol and shrinking the source probability linearly according to desired code rate $\frac{7}{11}$.

Symbol	Probability	Cumulative probability
f_1	0.5	0.5
a_1	0.25	0.75
a_2	0.125	0.875
a_3	0.0625	0.9375
a_4	0.0625	1.0

of shrinking the source probability. Suppose we want an arithmetic coder of source A with a per symbol code rate of R, i.e,

$$R = \frac{H(A)}{\overline{k'}}.$$

We have

$$\overline{k'} = \frac{H(A)}{R}.$$

On the other hand,

$$\overline{k'} = -\sum_i P(a_i) \log_2 P'(a_i).$$

We have

$$
\begin{aligned}
-\sum_{i} P(a_i) \log_2 P'(a_i) &= \frac{H(A)}{R} \\
&= -\frac{\sum_i P(a_i) \log_2 P(a_i)}{R} \\
&= -\sum_{i} P(a_i)(\frac{1}{R} \times \log_2 P(a_i)) \\
&= -\sum_{i} P(a_i) \log_2(P(a_i))^{\frac{1}{R}}.
\end{aligned}
$$

In this case, we can choose

$$
P'(a_i) = (P(a_i))^{\frac{1}{R}}
$$

To achieve an arithmetic code with rate R, the reserved probability space for forbidden symbols will be

$$
\epsilon = 1 - \sum (P(a_i))^{\frac{1}{R}}. \tag{3.2}
$$

The average encoded output length for each particular symbol a_i will be $-\frac{1}{R} \log_2 P(a_i)$, and the

Table 3.3: The probability table for encoding a source with four symbols after inserting a forbidden symbol and shrinking the source probability exponentially according to desired code rate $\frac{7}{11}$.

Symbol	Probability	Cumulative probability
f_1	0.4741	0.4741
a_1	0.3365	0.8106
a_2	0.1132	0.9238
a_3	0.0381	0.9619
a_4	0.0381	1.0

extra bits for each source symbol a_i will be $-\frac{1-R}{R} \log_2 P(a_i)$ (a quantity that is related to the source probability of a_i).

3.1.2   ERROR DETECTION CAPABILITY

The reserved probability space can be used for error detection. It has been shown that arithmetic codes with forbidden symbols are very effective for error detection. However, precisely calculating the probability that arithmetic codes can detect errors is rather complicated. Typically, the probability of detecting errors, $P_{detected}$, is calculated as $1 - P_{undetected}$ where $P_{undetected}$ is the probability that the error results in the decoding of another valid codeword. We can calculate this probability by determining the distance spectrum of arithmetic codes. Since arithmetic codes are variable length codes and nonlinear (as we will discuss later), the calculation of the distance spectrum is quite challenging. However, Chou and Ramchandran [19] gave an empirical model to estimate the number of bits necessary to detect an error after it occurs. The probability of not detecting an error after n bits is

$$P(n) = (1 - \epsilon)^n.$$

This probability becomes exponentially small as the number of decoded symbols grows.

3.1.3   ERROR CORRECTION WITH ARITHMETIC CODES

The reserved probability space in the source alphabet results in redundancy in the encoder output. This redundancy can also be used for error correction. Arithmetic codes can be viewed as tree codes. For example, the arithmetic encoding example from Chapter 2 can be described as tree coder as shown in Figure 3.2.

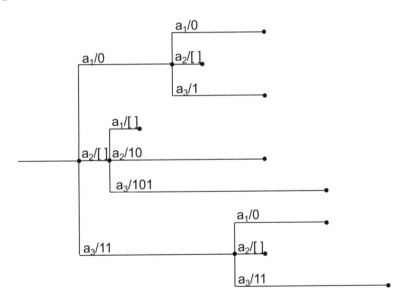

Figure 3.2: Arithmetic codes viewed as tree codes.

Sequential decoding is a general decoding algorithm for tree codes. It was introduced by Wozencraft and Reiffen to decode convolutional codes in [104]. Fano [29] presented an improved sequential algorithm in 1963, which is now known as the Fano algorithm. Zigangirov [108] and Jelinek [50] independently proposed a fast sequential algorithm, which is now known as the stack algorithm. The Fano algorithm performs forward and backward searches throughout the tree using

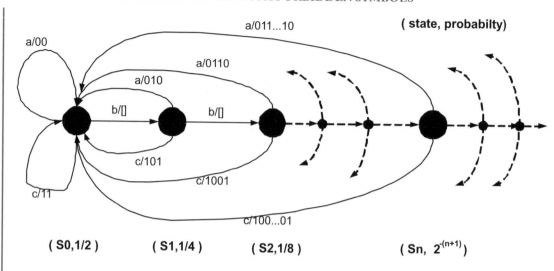

Figure 3.3: State transition diagram for Example 3.1 with 3 letter source and an unbounded *Scale3* value.

a threshold and partial path metric. One of the drawbacks of Fano algorithm is that the decoder moves only one node (forward or backward) per iteration, and nodes are often revisited and their partial path metrics are recomputed. In the stack algorithm, no node is visited more than once, and the nodes with the best partial path metrics are stored in memory and extended. Pettijohn, Hoffman, and Sayood [75, 74] proposed two sequential decoding algorithms, depth first and breadth first, for decoding arithmetic codes in the presence of channel errors.

3.1.3.1 Depth first algorithm

The depth first algorithm in [75, 74] is based on the Fano algorithm. In order to reduce the computational complexity, channel reliability information is used to classify the nodes into different categories. Given binary phase shift keying (BPSK) signaling over an AWGN channel, the region within $\pm\Delta$ around the origin is denoted as *Null zone* [31]. The received signals that fall into the *Null zone* region will be the branch nodes visited by the decoder. The value of Δ can be adaptively chosen to control the number of branch points. Although the reserved probability space alone can be used to detect errors and terminate the incorrect path, this approach is computationally expensive. Both Hamming distance and Euclidean distance are also used to prune paths. For Hamming distance, the partial path metric, K_h, is the number of corrections, and the threshold T_h is the maximum allowed Hamming distance between the hard decisions for the received sequence and the sequential decoder output sequence. For Euclidean distance, the partial path metric, K_c, at time n is defined as

$$K_c = \frac{1}{n} \sum_{k=1}^{n} (r_k - \widehat{x}_k)^2,$$

while the threshold T_c is defined as

$$T_c = \frac{1+\alpha}{N} \sum_{k=1}^{N} (r_k - \widetilde{x}_k)^2,$$

where r_k is the soft output of the channel, $\widehat{x}_k$ is the sequential decoder output, $\widetilde{x}_k$ is the hard decision for the received sequence, N is the sequence length and α is an experimentally determined offset.

3.1.3.2 Breadth First Algorithm

The breadth first algorithm in [75, 74] is a variant of the stack algorithm. The width of the *Null zone* 2Δ is fixed here. However, Δ needs to be chosen large enough to contain most errors, i.e, the probability that an error falls out of *Null zone* is small enough that we can ignore it. On the other hand, Δ cannot be so large that it will result in too many branch nodes and higher computation complexity. A trade off can be made based on the channel condition.

Another way to control the decoding complexity is to adaptively change the number of best paths kept in the decoder. The decoder first decodes the received sequence with a small stack size. If the arithmetic decoder can not find a sequence that decodes without errors in the stack, the size of the stack is increased, and the decoder tries to decode it again.

3.2   VIEWING ARITHMETIC CODES AS FIXED TRELLIS CODES

Arithmetic codes were modeled as state machines (or quasi-arithmetic codes) with the lower and upper limits as states to avoid multiplication and to achieve fast encoding and decoding in [46, 47]. The number of possible states depends on the register precision used for arithmetic coding. If an integer interval $[0, N-1]$ is used to replace the unit interval in the implementation of arithmetic coding, the maximum number of possible lower and upper limit pairs will be $3N^2/16$.[1] However, the branch outputs associated with each state may vary due to the E_3 condition and the *Scale3* counter. Here we use $\{l, u, Scale3\}$ [38] as the internal state of the arithmetic coder.

Example 3.1   This can lead to an infinite number of states as, in theory, *Scale3* is not bounded. For example, let's encode the following source $\{a, b, c\}$ with probabilities $\{1/4, 1/2, 1/4\}$ and cumulative probability $\{0, 1/4, 3/4, 1\}$ with fixed precision (5 bit) arithmetic codes. The encoding process is depicted in Table 3.4.

---

[1]This is calculated by counting all valid lower and upper limit pairs.

Table 3.4: Encoding process for Example 3.1 with a 3 letter source and an unbounded *scale3* value.

Current state	Input	Next state	Output
	a	$\{0, 31, 0\}$	00
$\{0, 31, 0\}$	b	$\{0, 31, 1\}$	[ ]
	c	$\{0, 31, 0\}$	11
	a	$\{0, 31, 0\}$	010
$\{0, 31, 1\}$	b	$\{0, 31, 2\}$	[ ]
	c	$\{0, 31, 0\}$	101
. . .	. . .	. . .	. . .

In practice, states with a large *Scale*3 are very low probability events. If we limit the maximum value of *Scale*3 allowed by forcing the encoder to generate an output and flush *Scale*3 when a threshold is reached, we introduce a very modest increase on the length of the encoded output[2] while resolving the issue of an infinite number of states. This idea was used in the IBM Q-coder to control the carry over problem [58]. If *Scale*3 is limited to a maximum value, then the aforementioned finite precision arithmetic coder yields a *finite* state machine that can be used to derive identical encoding and decoding trellises. Here we use an example to demonstrate how to obtain all of the states and state transitions.

Example 3.2   Suppose we have a binary source with alphabet $\{a, b\}$ with probabilities $\{3/4, 1/4\}$, respectively. We reserve a probability space of $1/5$ for a forbidden symbol f and reduce the space for the other symbols accordingly. Thus, we have (for encoding purposes) cumulative probabilities $\{0, 3/5, 4/5, 1\}$. We will use 5 bit fixed precision arithmetic coding with the maximum *Scale*3 value set to 3. The states of the above arithmetic coder can be found recursively. Recall that the state is given by the triple $\{l, u, Scale3\}$. The initial state is $\{0, 31, 0\}$, and, using all possible source inputs, new states are reached by the arithmetic encoder. Starting from the new states and using all possible source inputs, the arithmetic encoder either goes back to some known state or goes to a new state. Using this exhaustive search approach, we can determine that the above arithmetic code has 28 unique states in the original definitions as shown in Table 3.5. At this point, the arithmetic code can be viewed as a fixed-input-length variable-output-length trellis code.

DECODING ON A REDUCED STATE TRELLIS

If we use the Table 3.5 to build a trellis, there are many states with only one branch input. When applying the Viterbi decoding algorithm, there are no branch prunes at these states. Therefore, these states can be treated as internal states. The decoder knows what the previous state is at these points

---

[2]An example of this is shown using the modified arithmetic codes described in Table 3.6.

Table 3.5: Exhaustive state definitions for Example 3.2 source encoder.

State index	Current state	Input	Next state,	State index	Output
0	{0, 31, 0}	a	{0, 18, 0},	1	[ ]
		b	{4, 31, 0},	2	11
1	{0, 18, 0}	a	{0, 21, 0},	3	0
		b	{12, 27, 2},	4	[ ]
2	{4, 31, 0}	a	{4, 19, 0},	5	[ ]
		b	{8, 31, 0},	6	11
3	{0, 21, 0}	a	{0, 25, 0},	7	[ ]
		b	{4, 23, 0},	8	10
4	{12, 27, 2}	a	{8, 25, 3},	9	[ ]
		b	{0, 31, 0},	0	10010
...	...	...	...	...	...
25	{8, 23, 2}	a	{0, 17, 3},	27	[ ]
		b	{0, 31, 0},	0	01001
26	{12, 27, 4}	a	{0, 19, 0},	24	100000
		b	{0, 31, 0},	0	1000010
27	{0, 17, 3}	a	{0, 19, 0},	24	0111
		b	{0, 31, 0},	0	111111

when a path is traced back from the end of the sequence. Hence, the only states we need in the decoding trellis are those states with multiple inputs. We call those states reduced states, and we need to find all the transitions between the reduced states. The transitions between the reduced states can be found by the following procedure:

1. Select a reduced state as root node. The states connected to this state are leaf nodes.

2. If a leaf node is a reduced state, we do not need to do anything. If a leaf node is an internal state, we extend the tree to next level.

3. The extension is stopped if all the leaf nodes are reduced states.

4. Each path from the root node to a leaf node is a branch transition between reduced states. The concatenation of the inputs and outputs along the path forms the branch label for each transition.

The above procedure as applied to states 0 and 2 in the example described in Table 3.5 as demonstrated in Figure 3.4. The complete transitions between the reduced states using reduced state definitions are shown in Figure 3.5. It can be seen that the transitions between reduced states form a variable-input-length, variable-output-length trellis code. In this example, the number of states is

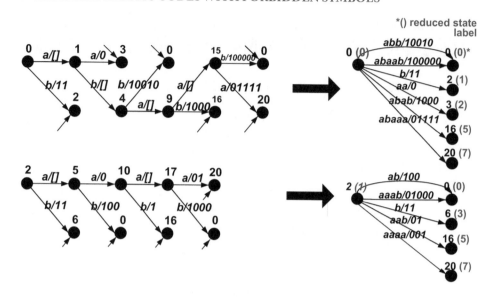

Figure 3.4: Finding transitions between reduced states using Example 3.2.

reduced from 28 to 10, but this reduction also results in multiple exit paths and parallel transitions in the trellis.

3.2.1   ENCODING

Encoding can be done either by performing ordinary arithmetic coding with a cap on *Scale*3 or using the state transitions to encode via table lookup. The decoder will use a trellis with the reduced states. Since it is possible that the input sequence ends at an internal state, the encoder needs to handle this eventuality. If ordinary arithmetic encoding is used, the encoder needs to check the final state information, and a termination to the reduced states is needed in case the final state is an internal state. For the table lookup method, the encoder also needs to encode additional symbols to generate the final output bits.

3.2.2   DECODING

Trellis based arithmetic codes belong to the set of variable length codes. Many methods have been proposed for decoding trellis based variable length codes (most focus on Huffman codes) [23], [90], [71], [72], [68], [8], [9], [24], [42]. Most of these methods can be directly used in our case. Here we use the popular list Viterbi algorithm combining the error detection capability of the modified arithmetic code with a 16 bit CRC checksum added at the end of each packet to detect small Hamming distance errors. The difference of our work from [24] lies in the fact that our decoding trellis arises naturally out of the state machine interpretation of arithmetic codes. No

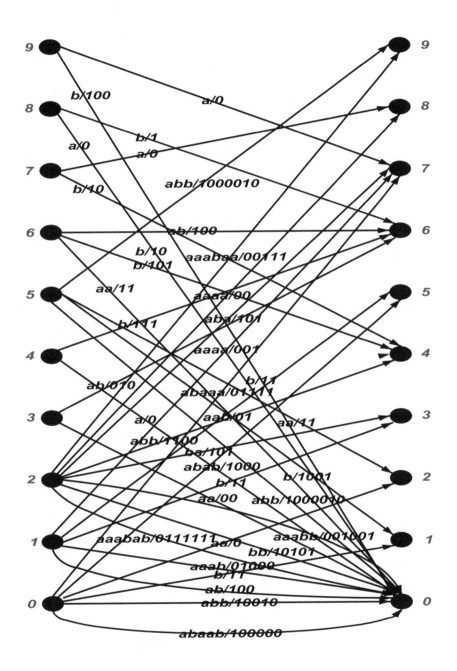

Figure 3.5: Transitions between reduced states for Example 3.2.

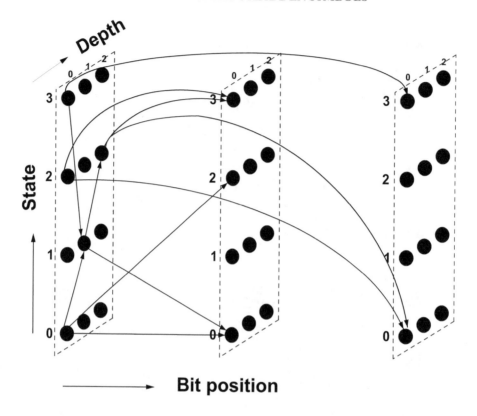

Figure 3.6: Decoding trellis - the three dimensions correspond to "time" (or bit position), state, and depth.

additional channel coder or concatenated scheme is used. We use the list Viterbi algorithm on the bit-constrained trellis. In order to deal with state transitions without any output, the decoder uses the list Viterbi algorithm on a three-dimensional trellis as shown in Figure 3.6. The three dimensions of the trellis are "time" (or bit position), state, and depth. The first dimension represents the bit position on the received sequence, the second dimension represents all possible states, and the third dimension captures the possible non-output state transitions. Note that we have chosen to work on the reduced trellis. It would also be possible to work directly on the original trellis.

3.3   SIMULATIONS WITH AN IID SOURCE

In order to compare with previous results from [105], a similar setting to that used in [105] is implemented (i.e., we use the same arithmetic codes (except for no *Scale3* limitation in [105]), but we use a different decoding algorithm, and no error correction code is added except for inserting the forbidden symbols and checksum in both papers). The source packet is 256 bits long. The forbidden symbol reserved probability space is chosen to make the overall channel coding rates close to 2/3, 4/5, and 7/8, respectively. The overall rate takes into account the CRC checksum, the limitation on $Scale3$ and the introduced forbidden symbol. The precision of the arithmetic coder is eight bits and the limitation of maximum $Scale3$ values are either 3, 4, or no limit. The source distribution, code rate, and other related information about the codes are listed in Table 3.6. The maximum depth is the maximum number of transitions without any output. It can be seen that the number of reduced states varies as we change the gap size, and the number of states used in the decoding trellis can be reduced roughly by a factor of 3 using the previously described state reduction process. The additional redundancy impact of different limitations on $Scale3$ is very small on such short sequences (as can be seen in Table 3.6). If no $Scale3$ limit is imposed on these case, the information rates increase by 0.004, 0.006, and 0.006, respectively, – at the cost of having a more complex coding trellis. The higher $Scale3$ limitation of 4, generally, results in a slight increase on the number of reduced states, but it does not have a large effect. In the following simulations, we used the codes generated with $Scale3$ limitation of 3.

The packet error rate results for BPSK signaling over an additive white Gaussian noise (AWGN) channel are shown in Figure 3.7. Both hard and soft decision list Viterbi decoding (list size =10) packet error rates are shown in the figure along with those from [105] for comparison. The packet error rates are plotted versus E_b/N_o in dB where E_b denotes the energy per information bit. For the lowest rate code (with gap 0.3), the performance with hard decision list Viterbi decoding already outperforms the iterative soft decoding scheme in [105]. In general, the performance of hard decision list Viterbi decoding is similar to the iterative decoding algorithm in [105] while the soft decision list Viterbi decoding algorithm provides much better performance. There is about a $2dB$ gain in E_b/N_o compared to the iterative decoding results at a packet error rate of 10^{-3}. Notice that unlike joint source channel coding systems, which rely mainly on residual redundancy in the source for error protection, schemes that include additional structure (such as gaps), show performance improvements at low error rates.

We also compared the performance of the list Viterbi decoder with the traditional separated schemes where the arithmetic code is protected by a convolutional code. Two separated schemes are used for comparison. First, an iterative arithmetic code with inner convolutional codes. The inner convolutional codes are chosen from [22] with memory of 5. A random interleaver is used between the inner codes and outer codes. For Soft AC decoding, a trellis is built according to [40] and [60] where the length constraint T is chosen to be 15 to reduce the trellis complexity and maintain the performance. For the second separated scheme, convolutional codes with similar rates (2/3, 4/5, and 7/8) with memory 8 (i.e., with comparable trellis complexity) were taken from [106]. Soft decision decoding is employed for convolutional codes followed by hard arithmetic decoding. The simulation stops after either reaching 100 packet errors or receiving 10, 000 packets. The packet error rates for these approaches on an AWGN channel are shown in Figure 3.8. For the rate 2/3 codes, the arithmetic code with list Viterbi decoding has about a $2dB$ gain over the corresponding convolutional

Table 3.6: Code parameters.

Probability frequency [a f b]	[20 15 15]			[20 6 15]			[20 3 15]		
Gap size	0.3			0.146			0.08		
Maximum Scale3	3	4	no limit	3	4	no limit	3	4	no limit
Average output packet length	410.5	409.8	408.4	332.6	331.3	330.1	302.4	301.3	300.2
Rate	0.633	0.634	0.637	0.782	0.785	0.788	0.86	0.863	0.866
Number of states	757	784	985	2839	3144	5063	3161	3510	5952
Number of reduced states	256	253	256	879	904	932	999	1032	1102
Maximum depth	1	1	1	3	4	4	3	4	4

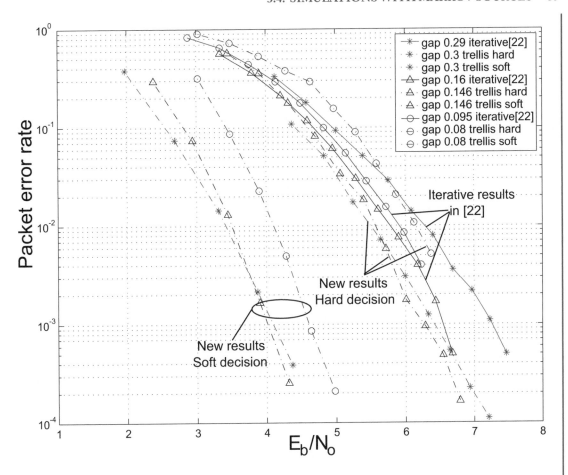

Figure 3.7: Simulation results—packet error rates versus E_b/N_o in dB.

code and a $0.7dB$ gain against the iterative approach at a packet error rate of 10^{-2}. While for the rate 4/5 and 7/8 codes, about $2.5dB$ and $3dB$ gains were obtained, respectively, over the corresponding convolutional codes. Compared with the iterative approach, our results have $1.2dB$ and $1.5dB$ gains, respectively. This shows that our approach is more attractive at higher rates.

3.4 SIMULATIONS WITH MARKOV SOURCES

For many sources, the source output symbols are correlated, and they can be modeled as Markov processes. Such sources can be compressed using context based arithmetic codes where different probability mapping functions are used for encoding based on past input symbols. In this paper, we will consider a first order Markov source with the transition probabilities $P(0|0) = \alpha$ and $P(1|1) = $

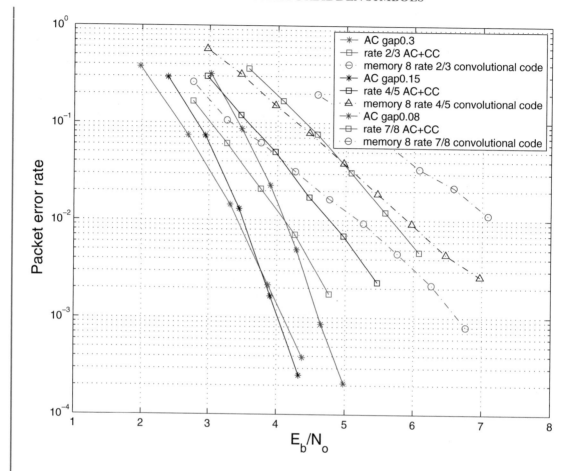

Figure 3.8: Comparison with conventional AC and convolutional coding and Joint AC and convolutional coding—packet error rates versus E_b/N_o in dB.

β, respectively. The source correlation ρ is defined as $\alpha + \beta - 1$. The steady state probabilities are $P(0) = \frac{1-\beta}{1-\rho}$ and $P(1) = \frac{1-\alpha}{1-\rho}$. The entropy rate of the Markov source is the expected value of the entropy over each state.

From the joint source and channel coding point of view, we can treat the Markov source in different ways. We could ignore the Markov structure in both the encoder and decoder. We can also keep some structure in the source coder output by treating the first order Markov source as an iid source but making use of this redundancy in the decoder. Finally, we can reserve some probability space in the conditional probability space to enable error detection and correction.

Here we will consider the following encoding and decoding configurations as shown in Figure 3.9. In scenario (a), although the source is a first order Markov source, both the encoder and

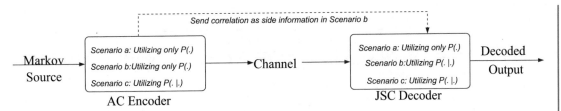

Figure 3.9: Three conceptual encoding/decoding diagrams for Markov sources.

decoder treat the source as an independent source and do not make use of the conditional probabilities. In scenario (b), the encoder treats the source as an independent source but sends the correlation parameter to the decoder as side information (note that this could also be estimated in a backwards adaptive manner at the decoder). The decoder makes use of the first order conditional probabilities when calculating the sequence prior probabilities. This scenario is useful for modeling non-stationary Markov sources where the correlation slowly changes over the packets. In scenario (c), both the encoder and decoder make use of the first order probabilities.

The decoding trellis used in scenario (a) is constructed based on the zero order probability of the Markov source. The error protection capability is provided by reserving space in the zero order probability space for the forbidden symbols. The prior probabilities of a sequence on the decoding trellis are calculated as the product of the zero order probabilities of the symbols in the sequence. The decoding trellis used in scenario (b) is the same as that used in scenario (a), but the prior probabilities of a sequence in the trellis are based on the first order conditional probability. In scenario (c), the trellis is generated using the first order probabilities of the Markov source. A probability space is reserved in the first order probability space for each source state to provide error protection.

We simulate the above systems using first order Markov sources with $P(0) = 4/7, P(1) = 3/7$ and correlation $\rho = 1/4, 1/3, 1/2, 2/3$ and $3/4$. We first look at scenarios (a) and (b), where the error correcting arithmetic codes with forbidden symbols are constructed based on the zero order source probabilities. For our example, we used the arithmetic codes used in our section 3.3 simulations where a gap size of 0.146 is inserted in the middle (middle column in Table 3.6). The overall rate R of the codes, i.e, the ratio of the first order entropy to the average output packet length, is listed in Table 3.7.

3.4.1   COMPARING SCENARIO (a) AND (b)

The simulation results for the AWGN channel using the list Viterbi decoding algorithm (list size 10) are shown in Figure 3.10 as a function of E_b/N_o. Notice that the encoders in scenario (a) and (b) are exactly the same and so are the decoding trellises. When the decoder makes use of the remaining redundancy in the source output, it can achieve performance gains. The decoders in scenario (b) have gains of $0.15, 0.4, 0.9, 1.5$ and $1.7 dB$ over the decoders in scenario (a) for Markov sources with $\rho = 1/4, 1/3, 1/2, 2/3$ and $3/4$, respectively.

Table 3.7: Code parameters for arithmetic codes.

scenario (a) and (b)					
ρ	$\frac{1}{4}$	$\frac{1}{3}$	$\frac{1}{2}$	$\frac{2}{3}$	$\frac{3}{4}$
Overall rate R	0.763	0.72	0.64	0.516	0.435
Average output packet length	333	333	333	333	333
scenario (c)					
ρ	$\frac{1}{4}$	$\frac{1}{3}$	$\frac{1}{2}$	$\frac{2}{3}$	$\frac{3}{4}$
Overall rate R	0.757	0.735	0.643	0.526	0.437
Average output packet length	335.5	326.7	331	327	331.8
Count_table [a f b]	0: [19 6 9] 1: [12 6 16]	0: [30 9 12] 1: [16 9 26]	0: [33 13 9] 1: [12 13 30]	0: [36 18 6] 1: [8 18 34]	0: [25 15 3] 1: [4 15 24]
Number of states	2343	1984	1324	772	533
Number of reduced states	726	627	393	243	165
Maximum Depth	4	4	3	3	2

3.4.2   COMPARING SCENARIO (b) AND (c)

For stationary Markov sources, context-based arithmetic codes can achieve better compression gains. This case is termed as scenario (c) in Figure 3.9. In such a situation, a probability space can also be reserved for forbidden symbols in the conditional probability space. For example, for the first order Markov source, forbidden symbols are inserted in both the $P(\cdot|0)$ and $P(\cdot|1)$ spaces. For the first order Markov sources described above, context-based arithmetic codes with forbidden symbols have been tested. The code parameters are listed in Table 3.7 where the gap is inserted such that the resulting context-based arithmetic codes have similar overall code rates.

The simulation results (dotted lines) are compared with those of scenario (b) (solid lines) as shown Figure 3.10. It can be seen that the decoder in scenario (c) has gains of 0.3, 0.2, 0.5, 1.0 and 1.2dB over the decoder in scenario (b) for Markov sources with $\rho = 1/4, 1/3, 1/2, 2/3$ and 3/4, respectively. It can be seen that the context-based arithmetic codes with forbidden symbols can further improve performance over the scenario (b) coders.

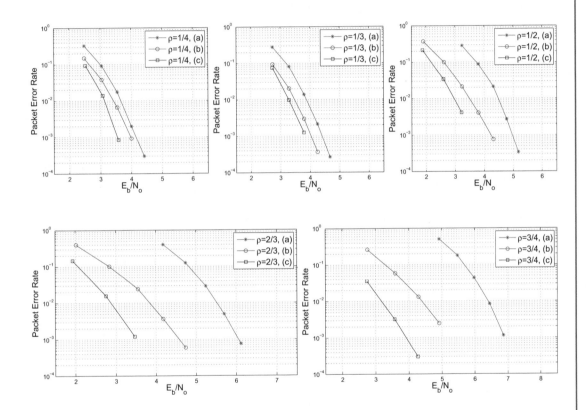

Figure 3.10: Simulation results for the three different scenarios. Each subplot corresponds to a different value of ρ.

CHAPTER 4

Distance Property and Code Construction

4.1 DISTANCE PROPERTY OF ARITHMETIC CODES

Based on the decoding trellis, we attempt to find the distance properties of codes generated by arithmetic codes with forbidden symbols. The trellis representation of the code gives us one way to explore its distance property. Before we focus on the distance property of arithmetic codes with forbidden symbols, we list two key differences between these codes and traditional error correcting codes, such as linear block codes and convolutional codes.

1. **In general, the codes are nonlinear.** This leads to the dependence of the weight distribution on the specific transmitted code sequence. The distance spectrum of the codes are thus the average of the weight distributions over the code sequences.

2. **The codes are variable length codes.** The branches in the trellis have different length.

Because of the nonlinearity, we cannot apply a fast search algorithm to compute the free distance and distance spectrum as can be done with convolutional codes [17]. Because of the variable code length transition, we cannot apply the concept of a product trellis to compute the distance of error events with the output transition matrix as can be done with trellis codes [13, 91]. Instead, we have to perform an exhaustive search over the trellis to determine the distance spectrum of arithmetic codes with forbidden symbols for all sequences of a specified length.

One of the reasons that people are interested in the distance properties of channel codes is that they can be used to derive error probability bounds. Therefore, we first review the error probability bound for trellis codes in order to determine what we need in terms of distance properties for variable length codes.

4.1.1 BOUND ON ERROR EVENTS

Let's look at the first error event probability bound on trellis codes with infinite sequence length [91]. The first error event probability $P_e(j)$, i.e., the probability that the decoder encounters its first error at time unit j, is given by

$$P_e(j) = \sum_c P(c) P(\bigcup_i e_i(c)|c), \qquad (4.1)$$

where $P(c)$ is the probability that the encoder chooses path c, and $e_i(c)$ is the ith error path that merges into the correct path c. Since a time invariant infinite trellis looks the same at every time unit, we can drop j and apply the union bound to obtain

$$P_e \leq \sum_c P(c) \sum_e P(e_i(c)|c). \tag{4.2}$$

We can rewrite the right side of Equation 4.2 in term of the path length as

$$P_e \leq \sum_{l \in \mathcal{L}} \frac{1}{P(l)} \sum_k P(c_{k,l}) \sum_i P(e_{i,k,l}|c_{k,l}), \tag{4.3}$$

where $\mathcal{L}$ is the set of valid path lengths, $P(c_{k,l})$ is the probability that the encoder chooses the kth path of length l, $e_{i,k,l}$ is the ith error path to path $c_{k,l}$, and $P(l)$ is the probability of all paths with length l. Note that for traditional fixed length trellis codes, the set $\mathcal{L}$ does not contain all the positive numbers. For example, the sequence length of the output of a $(2, 1, 2)$ convolutional encoder is always an even number. The set $\mathcal{L}$ contains all the even numbers and with a probability of $P(l) = 1$. For variable length codes, the set $\mathcal{L}$ usually contain all positive numbers, and $P(l)$ does not equal 1. For example, we can trace the sum of the probability of paths of different lengths for arithmetic codes with gap 0.3 from the initial state. The code parameters are listed in Table 4.1, and we choose register length $r = 8$. The probability is a function of the path length, and the probability of the path length is not one, and it stabilizes around 0.4 as path length increases. The path length gives the decoder extra information since we only need to select the path from a subset. For the distance spectrum calculation, we need to normalize the path probability with the path length probability.

If $BPSK$ modulation is used, $P(e_{i,k,l}|c_{k,l})$ is a function of d_m, the Hamming distance between the correct path and the error path. The upper bound can be rewritten as

$$P_e \leq \sum_{l \in \mathcal{L}} \sum_k \frac{P(c_{k,l})}{P(l)} \sum_m A_{m,k,l} f(d_m), \tag{4.4}$$

where $A_{m,k,l}$ is the number of error paths with Hamming distance d_m with respect to the correct path $c_{k,l}$, and $f(d_m)$ is the probability that decoder selects an error path with Hamming distance d_m from the correct path. Exchanging the order of the summation, we have

$$P_e \leq \sum_{l \in \mathcal{L}} \sum_m \sum_k \frac{P(c_{k,l})}{P(l)} A_{m,k,l} f(d_m) = \sum_{l \in \mathcal{L}} \sum_m A_{m,l} f(d_m), \tag{4.5}$$

where $A_{m,l}$ is the multiplicity, or the number, of error paths with Hamming distance d_m with respect to the correct path of length l, and

$$A_{m,l} = \sum_k \frac{P(c_{k,l}) A_{m,k,l}}{P(l)}. \tag{4.6}$$

4.1.2   USING THE BOUND TO GET ESTIMATE OF ERROR PROBABILITY

We typically only find the multiplicities up to certain path length, L_{max}. The above bound can then be broken into two components

$$P_e \leq \sum_{l \in \mathcal{L}, l \leq L_{max}} \sum_m A_{m,l} f(d_m) + \sum_{l \in \mathcal{L}, l > L_{max}} \sum_m A_{m,l} f(d_m) \tag{4.7}$$

For linear trellis codes, the second term can be bounded by the largest eigenvalue of the parallel transition matrix. For an arithmetic code with forbidden symbols, we do not have a bound. If we assume the second term is small enough, we can ignore the second term and get an estimate of the bound on the code performance, i.e,

$$P_e \approx \sum_{l \in \mathcal{L}, l \leq L_{max}} \sum_m A_{m,l} f(d_m). \tag{4.8}$$

4.1.3   DETERMINING THE MULTIPLICITY $A_{m,l}$

To determine the multiplicity $A_{m,l}$, we have to search through the trellis. In equation 4.6, $P(c_{k,l})$ is the probability of choosing the kth path with length l. In order to simplify this calculation, we can first determine all the possible paths of length l that terminate at node n and we have

$$A_{m,l} = \frac{1}{P(l)} \sum_k P(c_{k,l}) A_{m,k,l} = \frac{1}{P(l)} \sum_n P(n) \sum_s P(c_{s,l}|n) A_{m,s,n,l} = \frac{1}{P(l)} \sum_n P(n) A_{m,n,l}, \tag{4.9}$$

where $A_{m,s,n,l}$ is the number of error paths of length l with Hamming distance d_m with respect to a correct path $c_{s,l}$, and both the correct path and error path terminate at node n, $P(n)$ is the probability that the encoder selects state n at some time unit, and $p(c_{s,l}|n)$ is the probability that the decoder selects path s of length l, given that the path will terminate at node n. Based on the above equation, we can lay out the outline of our computing algorithm as follows:

Require: path length l, set $A_{m,l}$ to all zeros.
  for termination state $n = 0$ to maximum number of states **do**
    backward extend the trellis from state n up to length l;
    set $A_{m,n,l}$ to all zeros
    for path $s = 0$ to number of paths backward extended from state n **do**
      determine all the error paths that will not merge with the sth path until the termination
      state n
      for distance $m = 0$ to l **do**
        calculate the partial weight multiplicity $A_{m,s,n,l}$
      end for
      $A_{m,n,l} = A_{m,n,l} + P(c_{s,l}|n)A_{m,s,n,l}$, {weight by the conditional probability $P(c_{s,l}|n)$}.
    end for
    $A_{m,l} = A_{m,l} + P(n)A_{m,n,l}$, {weight by $P(n)$}.
  end for
  $A_{m,l} = \frac{1}{P(l)} A_{m,l}$, {weight by $1/P(l)$}.

The probabilities $P(n)$ and $P(l)$ can either be calculated based on the state transition diagram or estimated based on a Monte Carlo simulation. It can be seen that the above algorithm breaks down the computation of $A_{m,l}$ into calculating the partial multiplicity of given initial and ending states.

The brute force way to compute the distance profile of the given path is to visit all the valid paths and calculate the pair wise Hamming distance. Since the Hamming distance between two paths is additive and nondecreasing as the edges grow, we can compute the Hamming distances in a group from node to node and get the distance profile in one pass. More detailed calculations and other simplifications can be found in [11].

4.2   VERIFICATION

After determining the partial distance profile, we would like to compare our probability of error estimate with simulation results. For convolutional codes and trellis codes, the average source bit error probability is related to the first error event probability, and the estimate is compared with the simulated results. For arithmetic codes with forbidden symbols, different paths with the same length may correspond to different input symbol lengths. Even if no error has occurred after the first error event, the error may propagate due to the source symbol length mismatch. However, the output symbol error probability can be used to tie to the first error event to verify the estimate since the correct path and the error path have the same output length, and each error event will cause a certain number of output bit errors. Averaging for any time instant, the estimate of the output bit error probability P_{eo} is,

$$P_{eo} \approx \sum_{l \in \mathcal{L}, l \leq L_{max}}^{L_{max}} \sum_{m} \frac{m}{l} A_{m,l} f(d_m). \tag{4.10}$$

For an AWGN channel with BPSK signaling, the pairwise error probability of selecting the error path with Hamming distance d_m against the correct path will be

$$f(d_m) = Q\left(\sqrt{\frac{m \times E_s}{N_o}}\right). \tag{4.11}$$

where E_s is the signal energy per transmitted bit, N_o is the one-sided noise power spectral density and $Q(x)$ is the Q function [91]. Using (4.2) in the first error event estimate (4.1), we have

$$P_{eo} \approx \sum_{l \in \mathcal{L}, l \leq L_{max}}^{L_{max}} \sum_{m} \frac{m}{l} A_{m,l} Q\left(\sqrt{\frac{m \times E_s}{N_o}}\right). \tag{4.12}$$

Using the rate $R = 0.63$ arithmetic code with register length $r = 8$ listed in Table 4.1 to verify the above estimate, the partial distance spectra can be computed as shown in Figure 4.1. Based on

Table 4.1: Arithmetic codes with different register precisions.

Probability frequency [a f b]	[4 3 3]					
Register length r	4	5	6	7	8	9
Code rate R	0.619	0.639	0.634	0.631	0.634	0.634
Number of states	5	24	89	223	757	2749
Number of reduced states	3	6	30	75	256	914
Maximum depth	0	0	0	2	1	2
Average output length	420	407	410	412	410	410

computed partial distance spectra, the estimate on the output bit error rate over an AWGN channel can be calculated with Equation 4.12. The simulation result using soft Viterbi decoding algorithm over an AWGN channel is plotted with the estimates in Figure 4.2. The source block length is 256 and 100 packet errors are collected for each simulation point. It can be seen that the estimate is very accurate at high SNR. Also, it behaves as an upper bound at high SNR. In Figure 4.2, we also include the estimates for different maximum path lengths. It can be seen that the approximate bounds with different maximum path lengths diverge at lower SNR.

4.3   COMPLEXITY FACTORS AND FREEDOM IN THE CODE DESIGN

4.3.1   COMPLEXITY FACTORS

The number of possible lower and upper limit pairs of an arithmetic code is determined by the maximum integer number used to represent the unit interval, so the length of the register r of

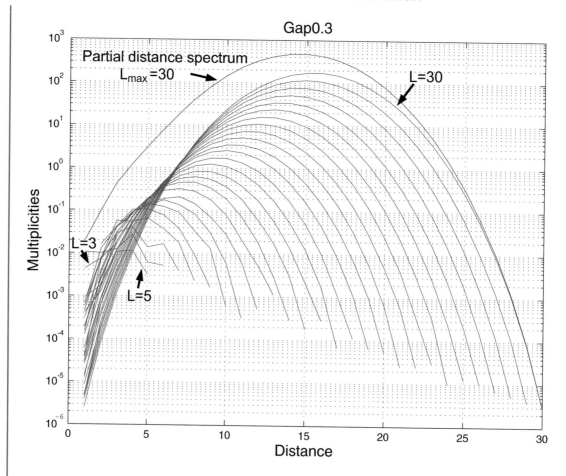

Figure 4.1: Partial distance spectrum ($L = 3 - 30$).

the arithmetic code is a major factor in the code complexity (The unit interval is represented by $[0\ r^2 - 1]$). In this chapter, we choose rate 0.63 arithmetic codes with forbidden symbols as examples, and, later, we use the constructed codes against the rate 2/3 turbo codes. Consider the rate 0.63 binary arithmetic code with one gap in the middle. The source is an independent and identically-distributed binary source with probability distribution specified in Table 4.1. We see a variety of codes obtained by changing the register length r.

The code rate and the average encoded output length of source with packet length 256 (and appended 16 bit CRC check sum) are also listed in Table 4.1. The above average output lengths are over 10, 000 packets. The code rate computation takes into account the appended CRC check sum, introduced gap and *scale3* limitations (i.e., using the average encoded output length of regular

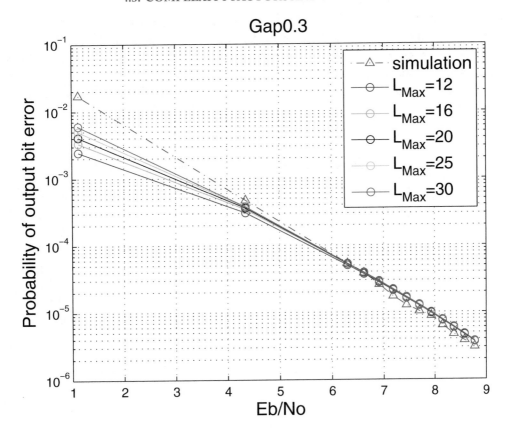

Figure 4.2: Comparison between simulation result and estimated error probabilities.

arithmetic codes against the above average output length). It can be seen that the higher computation precision does not always result in shorter output lengths. For this particular example, the shortest output packet length is achieved when $r = 5$. Also, the decoding complexity (the number of reduced states in the decoding trellis) increases significantly as the register length increases.

In general, the more states we have, the longer we can keep the paths separated and hence increase the Hamming distance between the paths. However, this is only true on the code ensemble. For a particular code, we need to investigate the distance spectrum of the code. Since the code performance at moderate to high SNR is determined by the first several components of the distance spectrum, we can expect better performance for codes with larger number of reduced states. However, the increasing register length does not guarantee a better code. The simulation results of these codes over an AWGN channel using the list Viterbi algorithm (list=10) are shown in Figure 4.3. 100 packet errors are conducted for each simulation point. Comparing the codes with $r = 6$ and $r = 7$, we obtain poorer performance for the $r = 7$ code. Analyzing the partial distance spectrum of these two codes, there is a larger Hamming distance 1 component in the distance spectrum of the $r = 7$ code (0.09 in $r = 7$ code versus 0.001 in $r = 6$ code as shown in Figure 4.6). It can be seen that the improvement of the code performance, if it exists, is not so significant as to warrant the increases in the decoding complexity. We will look at other ways to construct arithmetic codes with better performance.

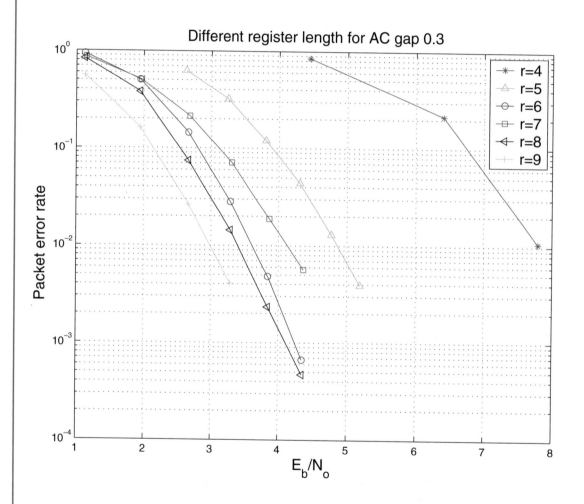

Figure 4.3: Packet error rate for codes with different register length.

4.3.2 FREEDOM IN THE CODE DESIGN

For the above rate 0.63 binary arithmetic code implemented with $r = 6$, different frequency table or probability mapping functions generate different arithmetic codes. For the probability mapping function, we first have the freedom to choose the order in which the symbols are placed.

4.3.2.1 Reordering the symbols

We have the freedom of arbitrarily reordering the symbols. This results in different arithmetic codes $(A - F)$ as listed in Table 4.2. Since there are three symbols in the probability space (including the forbidden symbol), permutation of these symbols results in six possible mapping functions. It can be seen that each mapping function generates a different code but they have a roughly similar complexity.

4.3.2.2 More forbidden symbols

We can also use more forbidden symbols to occupy the reserved probability space. We first use two forbidden symbols.

Two forbidden symbols

There are three positions that can be used for placing two forbidden symbols, i.e, at the beginning, in the middle, and at the end of the unit interval. The two forbidden symbols can occupy two out of the three positions. This results in six combinations $(G - L)$ as listed in Table 4.2 where f or f_i correspond to the forbidden symbols. We can also change the positions between the source symbols, the codes generated with these mapping functions $(M - R)$ are also listed in Table 4.2. So, there are a total twelve mapping functions with two forbidden symbols.

Three forbidden symbols

We can also use three forbidden symbols to occupy the three positions. Since the "Total" in the frequency count table is 10, each forbidden symbol frequency count is 1 for this example. We can only switch the source symbol to generate different codes $(S$ and $T)$. This results in the codes listed in S and T in Table 4.2.

We have total of 20 mapping functions under the constraint of the value of "Total" in the frequency count table. The number of reduced states for these arithmetic codes ranges from 22 to 53.

4.4 ARITHMETIC CODES WITH INPUT MEMORY

Different mapping functions result in different arithmetic codes. Although the source is a zero order source, we can go one step further to construct arithmetic codes using input memory as has been done in the construction of convolutional codes. The encoder will have a structure shown in Figure 4.4 where the past inputs are stored and are used to choose the mapping functions. The switching of

Table 4.2: Arithmetic codes with different mapping functions.

Index	A	B	C	D	E
Probability mapping function	[a f b]	[a b f]	[f a b]	[b f a]	[b a f]
Frequency count	[4 3 3]	[4 3 3]	[3 4 3]	[3 3 4]	[3 4 3]
Number of states /Maximum depth	89/0	71/1	124/1	85/1	143/1
Number of reduced states	30	27	43	31	42

Index	F	G	H	I	J
Probability mapping function	[f b a]	[a f_1 b f_2]	[a f_2 b f_1]	[f_1 a b f_2]	[f_2 a b f_1]
Frequency count	[3 3 4]	[4 2 3 1]	[4 1 3 2]	[2 4 3 1]	[1 4 3 2]
Number of states /Maximum depth	105/0	85/1	70/1	103/2	130/2
Number of reduced states	29	30	25	42	44

Index	K	L	M	N	O
Probability mapping function	[f_1 a f_2 b]	[f_2 a f_1 b]	[b f_1 a f_2]	[b f_2 a f_1]	[f_1 b a f_2]
Frequency count	[2 4 1 3]	[1 4 2 3]	[3 2 4 1]	[3 1 4 2]	[2 3 4 1]
Number of states/Maximum depth	98/1	101/2	73/1	90/1	146/1
Number of reduced states	29	32	22	30	41

Index	P	Q	R	S	T
Probability mapping function	[f_2 b a f_1]	[f_1 b f_2 a]	[f_2 b f_1 a]	[f_1 a f_2 b f_3]	[f_1 b f_2 a f_3]
Frequency count	[1 3 4 2]	[2 3 1 4]	[1 3 2 4]	[1 4 1 3 1]	[1 3 1 4 1]
Number of states/Maximum depth	146/1	90/1	90/1	124/1	153/2
Number of reduced states	52	25	30	47	53

the mapping function has influence only on how a particular interval is further partitioned. For example, if we assume that the initial past input is symbol a, the encoder partitions the unit interval with frequency table [a f b]; that is, the lower subinterval corresponds to the symbol a, the middle subinterval corresponds to the forbidden symbol, and the upper subinterval corresponds to the symbol b. If the first symbol is symbol a, the arithmetic encoder partitions the interval for the second input with frequency table [a f b]. If the first symbol is symbol b, the arithmetic encoder partitions the interval for the second input with frequency table [b f a]; that is, the lower subinterval corresponds to symbol b, the middle subinterval to the forbidden symbol and the upper subinterval to symbol a. Note, this has no influence on the encoding efficiency.

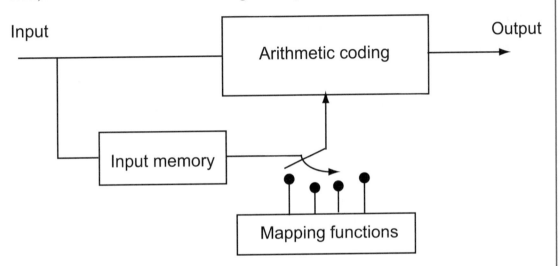

Figure 4.4: Arithmetic codes with memory.

4.4.1   MEMORY ONE ARITHMETIC CODES WITH FORBIDDEN SYMBOLS

We have 20 different mapping functions. So there are a total 400 of different combinations of memory one arithmetic codes with forbidden symbols. We use the code name to indicate what mapping functions are used and how they are used. For instance, the memory one code with a name DE uses the mapping function D when the past input is symbol a and utilizes the mapping function E when the past input is symbol b (Similarly, the memory two code with a name $DEFG$ will use the mapping function D, E, F and G when the past inputs are aa, ab, ba and bb, respectively). Note, that the memory one code AA is same as the memory zero code A. But memory one codes AB and BA are not identical. We calculated the partial distance spectra of memory one codes and singled out the best code TI based on these partial distance spectra. We compared the best code TI against codes A constructed with both $r = 6$ and $r = 8$ in Figure 4.7. It can be seen that the code TI has a gain of 0.5dB and 0.35dB, with respect to codes with $r = 6$ and $r = 8$, at a packet error rate of

10^{-3}, respectively. But code TI has a much smaller number of reduced states (77) when compared with code A with $r = 8$ (256).

4.4.2   MEMORY TWO ARITHMETIC CODES WITH FORBIDDEN SYMBOLS

The number of different possible combinations of probability functions grows exponentially as the length of input memory length, m, increases. For example, there are 20^4 possible different combinations of memory two arithmetic codes with forbidden symbols. For memory three arithmetic codes with forbidden symbols, the total number of combinations grows to 20^8. It becomes more and more difficult to compute all the partial distance spectra. Here we propose the following empirical method to search for good arithmetic codes with forbidden symbols with input memory m.

1. Set l equal to 1;

2. Select search parameter K where K^2 is the number of codes searched at each level l;

3. Select the top K memory l arithmetic codes with forbidden symbols that have a good partial distance spectrum.

4. Generate memory $l + 1$ arithmetic codes with forbidden symbols by combining two memory l arithmetic codes with forbidden symbols. This results in K^2 new codes.

5. $l = l + 1$ and if $l < m$ go to step 3;

6. Find the best memory m arithmetic codes with forbidden symbols.

In order to find out the proper value of the parameter K, we first let $K = 6$ and generate the memory two arithmetic codes based on the combinations of memory one arithmetic codes. A total of 36 memory two codes are generated. We found the best code to be $DCCM$. When $K = 12$, 144 memory two arithmetic codes are generated based on the combinations of memory one arithmetic codes. Code $CMOI$ has the best performance. Finally, we let $K = 24$ and generate 576 memory two arithmetic codes. We have the best memory two arithmetic code $DTTD$. The packet error rates for these codes we found with different value of K are shown in Figure 4.5. It can be seen that $K = 12$ is a good search parameter for this example when we trade off between the search effort and the performance of the codes we found.

4.4.3   MEMORY THREE ARITHMETIC CODES WITH FORBIDDEN SYM-BOLS

We built memory three arithmetic code with forbidden symbols based on the 12 selected memory two arithmetic codes, hence 144 memory three codes were constructed. We find the best memory three code $DTTDTDDC$ among the 144 codes. The code parameters together with those of the best memory two and memory one arithmetic codes are listed in Table 4.3. It can be seen that

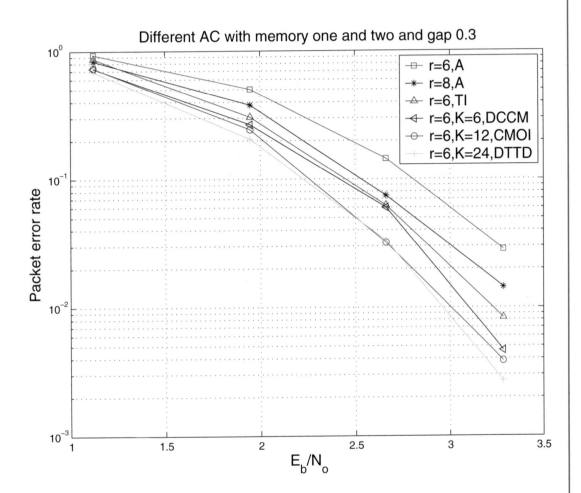

Figure 4.5: Packet error rates for the best codes for varying K (search parameter).

memory	0	0	1	2	3
register length	8	6	6	6	6
mapping functions	A	A	TI	$DTTD$	$DTTDTDDC$
Number of states	757	89	247	363	736
Number of reduced states	256	30	77	119	268
Maximum depth	1	0	1	1	1

Table 4.3: Parameters of arithmetic codes with memory.

the memory three arithmetic code with forbidden symbols and register length $r = 6$ has a similar complexity when compared with the arithmetic code A with forbidden symbols and register length $r = 8$. In Figure 4.6, we give the partial distance spectra for these codes. It can be seen that as we increase the memory length, the partial distance spectra of the selected best codes moves to the right. Their packet error rates are shown in Figure 4.7. The memory three arithmetic codes with forbidden symbols have a $0.8dB$ performance gain against code A with $r = 8$ when comparing at a packet error rate of 10^{-3}. In Figure 4.7, we also include the packet error rate simulation results of the rate 2/3 turbo codes [1] with packet length 256. The puncture patterns for information bits, first, parity bits and, second, parity bits are [1 1 1 1], [0 0 1 0] and [1 0 0 0], respectively. The number of iterations for the turbo code is 8. It can be seen that the performance of the integrated joint source and channel, arithmetic codes with memory two have performance close to a state of the art error correcting codes. The memory three code outperforms the turbo code by $0.5dB$ at a packet error rate of 10^{-3}.

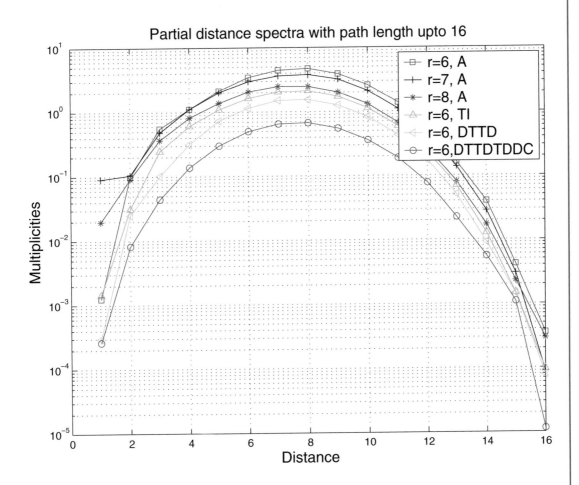

Figure 4.6: Partial distance spectra of arithmetic codes.

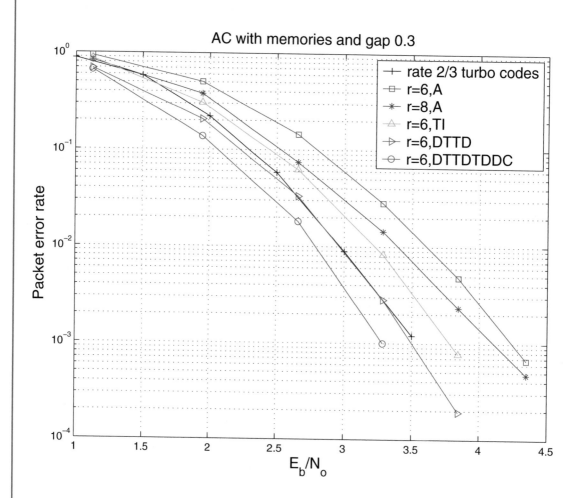

Figure 4.7: Packet error rates for arithmetic codes with memory.

CHAPTER 5

Conclusion

The arithmetic code along with being the quintessential lossless source coding technique also provides an excellent framework for implementing joint source channel coding approaches. The use of a reserved probability space for a "phantom" symbol permits the introduction of redundancy in a controlled fashion, which can be used for error protection. Interpreting the arithmetic code as a state machine both opens up new ways of understanding the error correction capability of the arithmetic codes with forbidden symbols, as well as providing ways to implement error correction algorithms.

In this monograph, we have tried to show different ways we can use a finite state machine interpretation of arithmetic codes with forbidden symbols. The number of states for an arithmetic encoder with finite precision implementation might be infinite. However, the limitation on the $Scale3$ variable in the encoding process will guarantee that the finite precision implementation of arithmetic coding is a finite state machine (although the limitation on $Scale3$ also results in a very small increase on the encoded sequence output length). The finite state machine interpretation leads to a fixed-input-symbol length and variable-output-bit length trellis with a fixed number of outgoing branches and variable number of incoming branches at each node. The nodes with one incoming branch can be treated as internal states since the Viterbi decoding algorithm stores one path at each node and hence no path pruning occurs at nodes with only one incoming branch. Thus, the trellis can be represented in a more compact way where only the nodes with multiple incoming branches are needed. With the state reduction process, the number of states used for decoding can be reduced by typically a factor of three. Finally, we can efficiently decode the source sequences with the list Viterbi algorithm on the trellis.

When the same arithmetic codes with forbidden symbols [105] are simulated at lower precision implementations, the simulation results in Figure 5.1 show the performance of a hard decision list Viterbi decoding is similar to the iterative decoding algorithm in [105] while the soft decision list Viterbi decoding algorithm provides much better performance. There is about a $2dB$ gain in E_b/N_o compared to the iterative decoding results at a packet error rate of 10^{-3}. Notice that unlike joint source and channel coding systems which rely mainly on residual redundancy in the source for error protection, schemes that include additional structure (such as gaps) show performance improvements at low error rates.

The finite state machine interpretation can be easily migrated to the Markov source cases. We can encode the Markov sources without considering the conditional probabilities, but the list Viterbi decoding algorithm can exploit the conditional probabilities. We can also use context-based arithmetic coding and apply the finite state machine interpretation on it. The decoding trellis derives from the context-based arithmetic coder and possesses both the characteristics of the Markov source and the arithmetic coding process. Hence, it provides better performance.

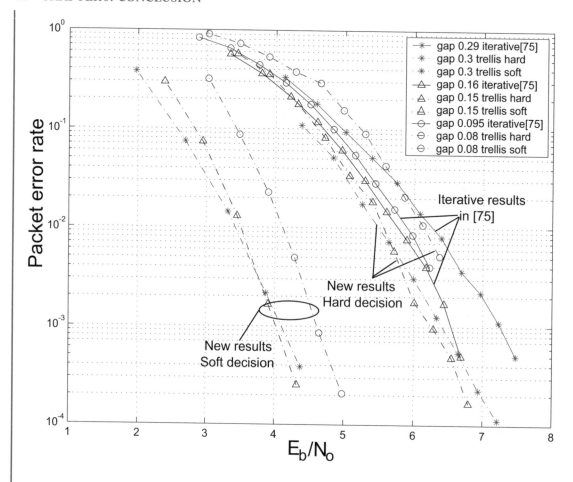

Figure 5.1:   Packet error rates for a variety of joint source and channel arithmetic coding approaches.

The finite state machine interpretation not only provides us better decoding performance, but it also allows us to more systematically understand the arithmetic codes. It allows us to estimate the partial distance spectra of the arithmetic codes and compare the codes based on their partial distance spectra estimates.

With this powerful tool in hand, we wanted to determine a way to design arithmetic codes with forbidden symbols such that the resulting joint source and channel coder can efficiently convert the source inputs into channel inputs for a given rate, and at the same time, we wanted the channel input sequences to have a good distance properties to overcome the channel noise. Based on these guidelines, we proposed arithmetic codes with memory. The proposed arithmetic coder encodes the source according to the cumulative probabilities, as in a standard arithmetic coder, but has the flexibility to arrange the source probabilities and inserted forbidden symbols. The introduced memory allows us to make use of this flexibility. Codes with good distance properties are searched for within the code parameter space. We also proposed an iterative way to construct and search high memory arithmetic codes based on low memory arithmetic codes to handle the code parameter space explosion problem that accompanies the increased memory. The memory length of arithmetic codes becomes a factor in the complexity of the decoding trellis. In contrast with the complexity increases due to increasing the implementation precision (where the complexity of decoding trellis increases by a factor of 4 when the implementation precision is increased by one bit), the complexity of the decoding trellis increases by a factor of 2 with one additional memory location. Using this observation, we can start with arithmetic codes with lower implementation precision to build higher memory arithmetic codes with forbidden symbols. This will result in arithmetic codes with the same rate and complexity but with better distance properties and better packet error rates compared with arithmetic codes constructed from higher implementation precision.

Bibliography

[1] 3G TS.25.212, "Universal Mobile Telecommunications System (UMTS); Multiplexing and channel coding (FDD)," 2000.

[2] N. Abramson, *Information Theory and Coding*, McGraw-Hill, 1963.

[3] R. Anand, K. Ramchandran, and I. Kozintsev, "Continuous error detection (CED) for reliable communication", *IEEE Trans. Commun.*, vol. 49, pp. 1540–1549, Sept. 2001. DOI: 10.1109/26.950341

[4] T. C. Ancheta, "Joint source channel coding", Ph.D. dissertation, Univ. Notre Dame, Aug. 1977.

[5] E. Ayanoğlu and R. M. Gray, "The design of joint source and channel trellis waveform coders", *IEEE Trans. Inform. Theory*, vol. IT-33, pp. 855–865, Nov. 1987.

[6] L. R. Bahl, J. Cocke, F. Jelinek, and J. Raviv, "Optimal decoding of linear codes for minimizing symbol error rate", *IEEE Trans. Inform. Theory*, vol. 20, pp. 284–287, Mar. 1974. DOI: 10.1109/TIT.1974.1055186

[7] B. A. Banister, B. Belzer and T. R. Fischer, "Robust image transmission using JPEG2000 and Turbo-codes", *IEEE Signal Processing Lett.*, vol. 9, pp. 117–119, Apr. 2002. DOI: 10.1109/97.1001646

[8] R. Bauer and J. Hagenauer, "Iterative source/channel-decoding using reversible variable length codes," in *IEEE DCC*, pp. 93–102, 2000. DOI: 10.1109/DCC.2000.838149

[9] R. Bauer and J. Hagenauer, "On variable length codes for iterative source/channel decoding," in *IEEE DCC*, pp. 273–282, 2001. DOI: 10.1109/DCC.2001.917158

[10] S. Ben-Jamaa, C. Weidmann, and M., Kieffer, "Analytical tools for optimizing the error correction performance of arithmetic codes," *IEEE Trans. Comm.*, pp.1458–1468, Sept. 2008. DOI: 10.1109/TCOMM.2008.060401

[11] D. Bi, "State machine interpretation of arithmetic codes for joint source and channel coding", *Ph.D. Dissertation*, Univ. of Nebraska at Lincoln, Jun. 2006.

[12] D. Bi, M. W. Hoffman and K. Sayood, "State Machine Interpretation of Arithmetic Codes for Joint Source and Channel Coding", *IEEE DCC*, pp. 143–152, 2006. DOI: 10.1109/DCC.2006.73

[13] E. Biglieri, "High-level modulation and coding for nonlinear satellite channels," *IEEE Trans. Commun.*, vol. 32, No. 5, pp. 616–626, May 1984. DOI: 10.1109/TCOM.1984.1096103

[14] C. Boyd, J. G. Cleary, S. A. Irvine, I. Rinsma-Melchert and I. H. Witten, "Integrating error detection into arithmetic coding", *IEEE Trans. Commun.*, vol. 45, pp. 1–3, Jan. 1997. DOI: 10.1109/26.554275

[15] V. Buttigieg, "Variable-length error-correcting codes", *Ph.D. Dissertation*, Univ. of Manchester, 1995.

[16] V. Buttigieg and P. G. Farrell, "Variable-length error-correcting codes", *Proc. IEE-Commun.*, vol. 147, pp. 211–215, Aug. 2000. DOI: 10.1049/ip-com:20000407

[17] M. Cedervall and R. Johannesson, "A fast algorithm for computing distance spectrum of convolutional codes," *IEEE Trans. Info. Theory*, vol. 35, No.6, pp. 1146–1159, Nov. 1989. DOI: 10.1109/18.45271

[18] K. Y. Chang and R. W. Donaldson, "Analysis, optimization and sensitivity study of differential PCM systems operating on noisy communication channels", *IEEE Trans. Commun.*, vol. COM-20, pp. 338–350, June 1972.

[19] J. Chou and K. Ramchandran, "Arithmetic coding-based continuous error detection for efficient ARQ-based image transmission", *IEEE Selected Areas in Commun.*, vol. 18, pp. 861–867, Jun. 2000. DOI: 10.1109/49.848240

[20] D. Comstock and J. D. Gibson, "Hamming coding of DCT compressed images over noisy channel", *IEEE Trans. Commun.*, vol. COM-32, pp. 856–861, July 1984. DOI: 10.1109/TCOM.1984.1096136

[21] R. V. Cox, J. Hagenauer, N. Seshadri, and C. E. Sundberg, "Subband speech coding and matched convolutional coding for mobile radio channels", *IEEE Trans. Signal Processing*, vol. 39, pp. 1717–1731, Aug. 1991. DOI: 10.1109/78.91143

[22] F. Daneshgaran, M. Laddomada and M. Mondin, "High-rate Recursive convolutional codes for concatenated channel codes," *IEEE Trans. Comm.*, pp. 1846–1850, Nov. 2004. DOI: 10.1109/TCOMM.2004.836590

[23] N. Demir and K. Sayood, "Joint Source/channel coding for variable length codes," in *IEEE DCC*, pp. 139–148, 1998.

[24] C. Demiroglu, M. W. Hoffman and K. Sayood, "Joint source/channel coding using arithmetic code and Trellis coded modulation," in *IEEE DCC.*, pp. 302–311, 2001. DOI: 10.1109/DCC.2001.917161

[25] J. G. Dunham and R. M. Gray, "Joint source and channel trellis encoding", *IEEE Trans. Inform. Theory*, vol. IT-27, pp. 516–519, July 1981 . DOI: 10.1109/TIT.1981.1056366

[26] G. F. Elmasry, "Arithmetic coding algorithm with embedded channel coding", *IEE Electron. Lett.*, pp. 1687–1688, Sept. 1997. DOI: 10.1049/el:19971145

[27] G. F. Elmasry, "Embedding channel coding in arithmetic coding", *IEE Proc.-Commun.*, vol. 146, pp. 73–78, Apr. 1999. DOI: 10.1049/ip-com:19990105

[28] ETS 300 401, "Digital Audio Broadcasting (DAB); DAB to mobile, portable and fixed receivers", 1997.

[29] R. M. Fano, "A heuristic discussion of probabilistic decoding", *IEEE Trans. Inform. Theory*, IT-9, pp. 64–74, Apr. 1963. DOI: 10.1109/TIT.1963.1057827

[30] N. Farvardin and V. Vaishampayan, "Optimal quantizer design for noisy channels: An approach to combined source-channel coding", *IEEE Trans. Inform. Theory*, vol. IT-33, pp. 827–838, Nov. 1987. DOI: 10.1109/TIT.1987.1057373

[31] R. D. Gitlin and E. Y. Ho, "Null zone decision feedback equalizer incorporating maximum likelihood bit detection", *IEEE Trans. Commun.*, vol. COM-23, pp. 1243–1250, Nov. 1975. DOI: 10.1109/TCOM.1975.1092744

[32] D. J. Goodman and C. E. Sundberg, "Combined source and channel coding for variable bit-rate speech transmission", *Bell Syst. Tech. J.*, vol. 62, pp. 2017–2036, Sept. 1983.

[33] D. J. Goodman and C. E. Sundberg, "Transmission errors and forward error correction in embedded differential PCM", *Bell Syst. Tech. J.*, vol. 62, pp. 2735–2764, Nov. 1983 .

[34] M. Grangetto and P. Cosman, "Map decoding of arithmetic codes with a forbidden symbol", in *Proc. Advanced Concepts for Intelligent Vision Systems, (ACIVS02)*, Sept. 2002.

[35] M. Grangetto, E. Magli, and G. Olmo, "A syntax-preserving error resilience tool for JPEG 2000 based on error correcting arithmetic coding," *IEEE Trans. Image Processing*, pp. 807–818, Apri. 2006. DOI: 10.1109/TIP.2005.863953

[36] M. Grangetto, B. Scanavino and G. Olmo, "Joint source-channel iterative decoding of arithmetic codes", *IEEE ICC,*, pp. 886–890, 2004.

[37] M. Guazzo, "A general minimum redundancy source coding algorithm", *IEEE Trans. Inform. Theory*, vol. 26, pp. 15–25, Jan. 1980. DOI: 10.1109/TIT.1980.1056143

[38] T. Guionnet and C. Guillemot, "Soft decoding and synchronization of arithmetic codes: application to image transmission over noisy channels", *IEEE Trans. Image Processing*, vol. 12, pp. 1599–1609, Dec. 2003. DOI: 10.1109/TIP.2003.819307

[39] T. Guionnet and C. Guillemot, "Joint source-channel decoding of quasi-arithmetic codes", *IEEE DCC*, pp. 272–281, 2004. DOI: 10.1109/DCC.2004.1281472

[40] T. Guionnet and C. Guillemot, "Soft and joint source-channel decoding of quasi-arithmetic codes", *EURASIP Journal on Applied Signal Processing*, pp. 393–411, 2004. DOI: 10.1155/S1110865704308085

[41] L. Guivarch, J-C. Carlach, and P. Siohan, "Joint source-channel soft decoding of Huffman codes with turbo codes", *IEEE DCC*, pp. 83–92, 2000. DOI: 10.1109/DCC.2000.838148

[42] A. Hedayat and A. Nosratinia, "List-decoding of variable-length codes with application in joint source-channel coding," *IEEE Asilomar Conf.*, pp. 21–25, 2002. DOI: 10.1109/ACSSC.2002.1197143

[43] E. hLogashanmugam, B. Sreejaa, R. Ramachandran, "Arithmetic Coding with Forbidden Symbol and Optimized Adaptive Context Tree Modeling (GRASP Algorithm) for H.264 Video Coders," *IEEE ICSCN*, pp. 599–602, Feb. 2007. DOI: 10.1109/ICSCN.2007.350679

[44] K. P. Ho and J. M. Kahn, "Transmission of analog signals using multicarrier modulation: A combined source-channel coding approach", *IEEE Trans. Commun.*, vol. 44, pp. 1432–1443, Nov. 1996. DOI: 10.1109/26.544460

[45] B. Hochwald and K. Zager, "Tradeoff between source and channel coding", *IEEE Trans. Info. Theory*, vol. 43, pp. 1412–1424, Sept. 1997.

[46] P. G. Howard and J. S. Vitter, "Practical implementations of arithmetic coding," in *Image and Text Compression*, J.A. Storer, ed., Kluwer Academic Publishers, 1992.

[47] P. G. Howard and J. S. Vitter, "Design and analysis of fast text compression based on quasi-arithmetic coding," *Inform. Proc. and Management*, vol. 30, No.6, pp. 777–790, Jun. 1994.

[48] M. Jeanne, J-C. Carlach, and P. Siohan, "Joint source-channel decoding of variable-length codes for convolutional codes and Turbo codes", *IEEE Trans. Commun.*, vol. 53, pp. 10–15, Jan. 2005. DOI: 10.1109/TCOMM.2004.840664

[49] F. Jelinek, *Probabilistic Information Theory*, McGraw Hill, 1968.

[50] F. Jelinek, "A fast sequential decoding algorithm using a stack", *IBM Journal of Research and Development*, vol. 13, pp. 675–685, Nov. 1969.

[51] I. Kozintsev, J. Chou, and K. Ramchandran, "Image transmission using arithmetic coding based continuous error detection", *IEEE DCC*, pp. 339–348, Mar. 1998. DOI: 10.1109/DCC.1998.672162

[52] S. I. Krich and T. Berger, "Coding for a delay-dependent fidelity criterion", *IEEE Trans. Inform. Theory*, vol. IT-20, No. 1, pp. 77–85, Jan. 1974.

[53] A. J. Kurtenbach and P. A. Wintz, "Quantizing for noisy channels", *IEEE Trans. Commun.*, vol. COM-17, pp. 291–302, Apr. 1969. DOI: 10.1109/TCOM.1969.1090091

[54] K. Lakovic and J. Villasenor, "On design of error-correcting reversible variable length codes", *IEEE Commun. Lett.*, vol. 6, pp. 337–339, Aug. 2002. DOI: 10.1109/LCOMM.2002.802041

[55] K. Lakovic, J. D. Villasenor, R. Wesel, "Robust joint Huffman and convolutional decoding," in *Proc. IEEE Vehicle Technology Conference*, pp. 2551–2555, 1999. DOI: 10.1109/VETECF.1999.800247

[56] C. Lamy and F-X. Bergot, "Lower bounds on the existence of binary error-correcting variable-length codes", in *Proc. IEEE Inform. Theory Workshop*, pp. 300–303, Mar. 2003.

[57] C. Lan, T. Chu, K. R. Narayanan, and Z. Xiong, "Scalable image and video transmission using irregular repeat-accumulate codes with fast algorithm for optimal unequal error protection", *IEEE Trans. Commmun.*, vol. 52, pp. 1092–1101, July, 2004. DOI: 10.1109/TCOMM.2004.831406

[58] G. G. Langdon and J. Rissanen, "Compression of black-white images with arithmetic coding," *IEEE Trans. Comm.*, vol. 29, No. 6, pp. 858–867, Jun. 1981.

[59] J. Liu and T. Li, "Iterative Joint Source Channel Decoding of Error Correction Arithmetic Codes," *IEEE Workshop on Signal Processing System*, pp. 346–350, Oct. 2007. DOI: 10.1109/SIPS.2007.4387570

[60] S. Malinowski, H. Jegou, and C. Guillemot, "Error recovery properties of quasi-arithmetic codes and soft decoding with length constraint," *IEEE ISIT*, Jul. 2006. DOI: 10.1155/2008/752840

[61] G. Martin, "Range encoding: an algorithm for removing redundancy from a digitized message", *Video and Data Recording Conf.*, England, Jul. 1979.

[62] J. L. Massey, "Joint source and channel coding", in *Communication systems and random process theory*, J. K. Skwirzynski, Ed. The Netherlands: Sijithoff and Nordhoff, pp.279–293, 1978.

[63] Asha Mehrotra, *GSM System Engineering*, Artech House, 1997.

[64] J. W. Modestino, V. Bhaskaran and J. B. Anderson, "Tree encoding of images in the presence of channel error", *IEEE Trans. Inform. Theory*, vol. IT-27, pp. 667–697, Nov. 1981. DOI: 10.1109/TIT.1981.1056428

[65] J. W. Modestino and D. G. Daut, "Combined source channel coding of image", *IEEE Trans. Commun.*, vol. COM-27, pp. 1644–1659, Nov. 1979. DOI: 10.1109/TCOM.1979.1094335

[66] J. W. Modestino, D. G. Daut and A. L. Vickers, "Combined source channel coding of images using the block cosine transform", *IEEE Trans. Commun.*, vol. COM-29, pp. 1262–1274, Sept. 1981. DOI: 10.1109/TCOM.1981.1095155

[67] C. C. Moore and J. D. Gibson, "Self-orthogonal convolutional coding for the DPCM-AQB speech encoding", *IEEE Trans. Commun.*, vol. COM-32, pp. 980–982, Aug. 1984. DOI: 10.1109/TCOM.1984.1096155

[68] A. H. Murad and T. E. Fuja, "Robust transmission of variable-length encoded sources", *in Proc. 1999 IEEE Wireless and Networking Conf.*, vol. 2, pp. 968–972, Sept. 1999. DOI: 10.1109/WCNC.1999.796815

[69] A. Nazer and F. Alajaji, "Unequal error protection and source-channel decoding of CELP speech", *IEE Elec. Lett.*, vol. 38, pp-347–349, Mar. 2002. DOI: 10.1049/el:20020242

[70] K. N. Ngan and R. Steele, "Enhancement of PCM and DPCM images corrupted by transmission errors", *IEEE Trans. Commun.*, vol. COM-30, pp. 257–269, Jan. 1982. DOI: 10.1109/TCOM.1982.1095366

[71] M. Park and D. J. Miller, "Decoding entropy-coded symbols over noisy channels by MAP sequence estimation for asynchronous HMMs", *in Proc. Conf. Information Sciences and Systems*, pp. 477–482, Mar. 1998.

[72] M. Park and D. J. Miller, "Joint source-channel decoding for variable-length encoded data by exact and approximate MAP sequence estimation", *IEEE Trans. Commun.*, vol. 48, pp. 1–6, Jan. 2000. DOI: 10.1109/ICASSP.1999.760626

[73] R. Pasco, *Source coding algorithms for fast data compression*, Stanford Univ., Ph.D. Dissertation, 1976.

[74] B. D. Pettijohn, M. W. Hoffman, and K. Sayood, "Joint source/channel coding using arithmetic code", *IEEE Trans. Commun.*, vol. 49, pp. 826–836, May 2001. DOI: 10.1109/26.923806

[75] B. D. Pettijohn, M. W. Hoffman and K. Sayood, "Joint source/channel coding using arithmetic code", *IEEE DCC.*, pp. 73–82, 2000. DOI: 10.1109/26.923806

[76] N. Phamdo and N. Farvardin, "Optimal detection of discrete markov sources over discrete memoryless channels - applications to combined source-channel coding", vol. 40, pp. 186–193, Jan. 1994.

[77] R. C. Reininger and J. D. Gibson, "Backward adaptive lattice and transversal predictors in ADPC", *IEEE Trans. Commun.*, vol. COM-33, pp. 74–82, Jan. 1985. DOI: 10.1109/TCOM.1985.1096199

[78] R. C. Reininger and J. D. Gibson, "Soft decision demodulation and transform coding", *IEEE Trans. Commun.*, vol. COM-31, pp. 572–577, Apr. 1983. DOI: 10.1109/TCOM.1983.1095837

[79] J. J. Rissanen, "Generalized Kraft inequality and arithmetic coding", *IBM J. Res. Develp.*, vol. 20, pp. 198–203, May 1976.

[80] J. Rissanen and G. G. Langdon, "Universal modeling and coding", *IEEE Inform. Theory*, vol. IT-27, pp. 12–23, Jan. 1981. DOI: 10.1109/TIT.1981.1056282

[81] J. J. Rissanen and G. G. Langdon, "Arithmetic coding", *IBM J. Res. Develop.* vol. 23, Mar. 1979.

[82] F. Rubin, "Arithmetic stream coding using fixed precision registers", *IEEE Trans. Inform. Theory*, vol. 25, pp. 672–675, Nov. 1979. DOI: 10.1109/TIT.1979.1056107

[83] A. Said, "Introduction to arithmetic coding - theory and practice", *Lossless Compression Handbook*, K. Sayood Ed., Academic Press, 2002.

[84] J. Sayir, *On coding by probability transformation*, Hartung-Gorre Verlag Konstanz, 1999.

[85] K. Sayood, *Introduction to Data Compression*, 2ed, Morgan Kaufmann Publishers, 2000.

[86] K. Sayood and J. C. Borkenhagen, "Use of residual redundancy in the design of joint source/channel coders", *IEEE Trans. Commun.*, vol. 39, pp. 838–846, Jun. 1991. DOI: 10.1109/26.87173

[87] K. Sayood and J. C. Borkenhagen, "Utilization of correlation in low rate DPCM systems for channel error protection", *in Proc. IEEE Int. Conf. Commun.*, pp. 1888–1892, June 1986.

[88] K. Sayood, F. Liu, and J. D. Gibson, "A joint source/channel coder design", *in Proc. IEEE ICC*, pp. 727–731, May 1993. DOI: 10.1109/ICC.1993.397369

[89] K. Sayood, F. Liu, and J. D. Gibson, "A constrained joint source/channel coder design", *IEEE J. Select. Areas Commun.*, vol. 12, pp. 1584–1593, Dec. 1994. DOI: 10.1109/49.339927

[90] K. Sayood, H. H. Otu, and N. Demir, "Joint source/channel coding for variable length codes", *IEEE Trans. Commun.*, vol. 48, pp. 787–794, May 2000. DOI: 10.1109/26.843191

[91] C. Schlegel, *Trellis Coding*, IEEE Press, 1997.

[92] C. E. Shannon, "A mathematical theory of communication", *Bell System Technical Journal* vol. 27, pp. 279–423, July, Oct. 1948.

[93] P. G. Sherwood and K. Zager, "Progressive image coding for noisy channels", *IEEE Signal Processing Lett.*, vol. 4, pp. 189–191, July 1997. DOI: 10.1109/97.596882

[94] A. Shiozaki, "Unequal error protection of PCM signals by self-orthogonal convolutional codes", *IEEE Trans. Commun.*, vol. 37, pp. 289–290, Mar. 1989. DOI: 10.1109/26.20103

[95] M. Srinivasan, R. Chellapa, and P. Burlina, "Adaptive source-channel subband video coding for wireless channels", *in Proc. 1997 Workshop on Multimedia Signal Processing*, June 1997. DOI: 10.1109/MMSP.1997.602669

[96] R. Steele, D. J. Goodman and C. A. McGonegal, "A difference detection and correction scheme for combatting DPCM transmission errors", *IEEE Trans. Commun.*, vol. COM-27, pp. 252–255, Jan. 1979.

[97] C. E. Sundberg, "The effect of single bit errors in standard nonlinear PCM systems", *IEEE Trans. Commun.*, vol. COM-24, pp. 1062–1064, June 1976. DOI: 10.1109/TCOM.1976.1093418

[98] V. A. Vaishampayan and N. Farvardin, "Optimal block cosine transform image coding for noisy channels", *IEEE Trans. Commun.*, vol. 38, pp. 327–336, Mar. 1990. DOI: 10.1109/26.48890

[99] J. Wang, L-L. Yang, and L. Hanzo, "Iterative construction of reversible variable-length codes and variable-length error-correcting codes", *IEEE Commun. Lett.*, pp. 671–673, Nov. 2004. DOI: 10.1109/LCOMM.2004.837645

[100] J. Wen and J. D. Villasenor, "Utilizing soft infomation in decoding of variable length codes", *IEEE DCC*, pp. 131–139, 1999.

[101] J. Wen and J. D. Villasenor, "Soft-input soft-output decoding of variable length codes", *IEEE Trans. Commun.*, vol. 50, pp. 689–692, May 2002. DOI: 10.1109/TCOMM.2002.1006449

[102] T. Wenisch, P. F. Swaszek, and A. K. Uht, "Combined error correcting and compressing codes", *IEEE ISIT*, pp. 238, 2001. DOI: 10.1109/ISIT.2001.936101

[103] I. H. Witten, R. M. Neal, and J. G. Cleary, "Arithmetic coding for data compression", *Commun. ACM*, vol. 30, pp. 520–540, Jun. 1987. DOI: 10.1145/214762.214771

[104] J. M. Wozencraft and B. Reiffen, *Sequential Decoding*, Cambridge, MIT Press, 1961.

[105] L. Xu, M. W. Hoffman and K. Sayood, "Hard decision and iterative joint source channel coding using arithmetic codes", *IEEE DCC*, pp. 203–212, 2005. DOI: 10.1109/DCC.2005.43

[106] Y. Yasuda, K. Kashiki, and Y. Hiratu, "High-rate punctured convolutional codes for soft decision Viterbi decoding," *IEEE Trans. on Commun.*, pp. 315–319, Mar. 1984. DOI: 10.1109/TCOM.1984.1096047

[107] M. Zhang, Y. Wang, J. Wang, S. Zhou, "A Hard Decision Error Correction Scheme for Corrupted Arithmetic Codes," *IEEE Conference on Computational Intelligence and Security*, pp. 344–346, Dec. 2007. DOI: 10.1109/CIS.2007.77

[108] K. Zigangirov, "Some sequential decoding procedures", *Probl. Peredachi Inf.*, vol. 2, pp. 13–25, 1966, Russian.